ABOUT THE AUTHOR

Alice Roberts is an anatomist and anthropologist, television presenter, author and Professor of Public Engagement with Science at the University of Birmingham. She has presented *Coast*, *Horizon* and several series about human evolution – including *The Incredible Human Journey*, *Origins of Us* and *Prehistoric Autopsy* – on BBC 2. She has also presented *Inside Science* on Radio 4, and writes a regular science column for *The Observer*. She lives near Bristol with her husband and two small children.

ALSO BY ALICE ROBERTS

Evolution: The Human Story

The Incredible Human Journey

The Complete Human Body

Don't Die Young:
An Anatomist's Guide to Organs and Your Health

THE INCREDIBLE UNLIKELINESS OF BEING

Evolution and the Making of Us

Alice Roberts

Illustrations by Alice Roberts

HERON

BOOKS

First published in Great Britain in 2014 by Heron Books
This edition published in Great Britain in 2015 by Heron Books
an imprint of

Quercus Publishing Ltd
Carmelite House
50 Victoria Embankment
London EC4Y 0DZ

An Hachette UK company

A CIP catalogue record for this book is available
from the British Library

PB ISBN 978 1 84866 479 1
EBOOK ISBN 978 1 84866 478 4

10 9 8 7 6 5

Typeset in Minion Pro by IDSUK (DataConnection) Ltd

Printed and bound in Great Britain by
Clays Ltd, St Ives, plc

In loving memory of Pam Stevens

CONTENTS

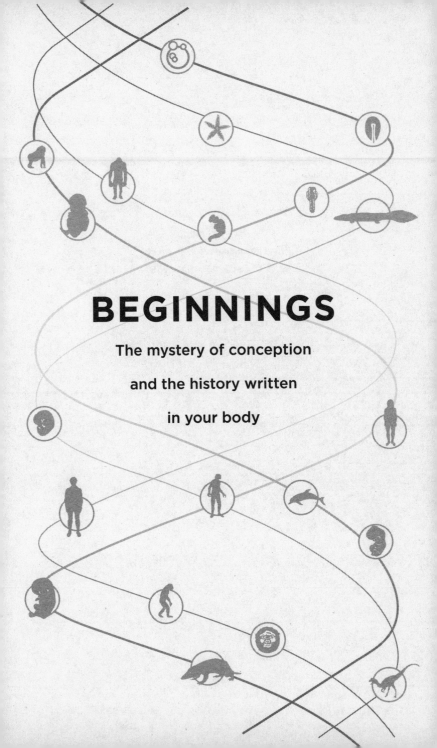

BEGINNINGS

The mystery of conception

and the history written

in your body

'Ex ovo Omnia'
('Everything comes from an egg')
WILLIAM HARVEY (1651)

The way I look at the world, and my concept of myself in time and space, has been completely transformed by becoming a mother. I gave birth to my first baby in 2010, and at that precise moment, I had an incredible, almost mystical, feeling of being connected – connected with my own ancestors, and connected with my descendants. I felt *more than* an individual: I was a link in a chain of life. For me, it was a very female thing. I was giving birth to a daughter, just as I had come from my mother, and she from her mother, and back and back in time.

If you're a man reading this, while you may not be able to give birth you could reflect in a similar way about your Y chromosome, which connects you to a lineage of male ancestors. Admittedly, it might not feel quite as epic, but then there's nothing like childbirth to make a thought seem momentous.

As an expectant mum in a twenty-first-century developed country, I had the incredible opportunity to see each of my babies *before* they were born. I remember the soaring joy I felt, just twelve weeks into my pregnancy, when I saw my tiny daughter for the first time, floating in her own small pond of amniotic fluid. At that point, I didn't even know she was a girl. It was wonderful to see her, but there was still a great gulf between that image on the screen and the experience of being pregnant.

I'm an anatomist – it's my business to know the structure of the human body and how it develops – but however much I knew about the development of an embryo, it didn't lessen that feeling that what was taking place inside me was utterly miraculous. Fertilisation is incredible in itself, and then the idea of a fertilised human egg, a single cell, transforming into something as complex as a complete human being is astonishing. By the time of my twelve-week scan, so much of the fetus had already been formed: it already had arms and legs, fingers and toes, guts and a beating heart. Already, it looked like a miniature baby. How did it get there, from just one cell at the moment of conception?

There's so much *unlikeliness* in you being here, right now, reading this. There's the unlikeliness of your parents meeting each other. There would have been so many points when their lives could have turned out differently, when each of them could have met someone else. Once they had hooked up, there's the unlikeliness of that ONE egg meeting that ONE sperm which made you. But I think that this unsettling feeling of the unlikeliness of being goes even further.

The development of a fertilised egg, a single cell, into a whole human being seems to defy belief. It appears to be some kind of biological miracle. But it's a miracle that doesn't require you to believe in any supernatural or divine intervention; it's a natural miracle, and over the last few centuries, scientists have unravelled many of the secrets of this incredible transformation (although there are certainly some secrets left to be discovered). At first glance, the development of a single egg into a whole person seems like such an impossible feat, such an *unlikely* occurrence, that we need to imagine some kind of supernatural guiding hand for this to happen, but when we understand the process in more detail, we can see how molecules, cells and tissues can build themselves into the organs of our bodies. It's a fundamental process which unites us with every other animal on the planet.

When you think about your own beginning, it's almost impossible to believe that you were once just a single cell: a fertilised egg, but you *know* that this is true. It may seem unlikely, but your very existence proves that it happened. You might also find it hard to believe that you could be descended from ancestors who, long ago, were also just single cells. But once you've come to terms with the undeniable fact that you yourself developed from a single cell during your embryonic development, perhaps it's easier to believe that you, that *we* as a species, have evolved from such humble beginnings. Looking to your more recent (but still, admittedly, quite ancient) ancestry, you find ancestors who were worms. You can trace your ancestry, moving along your own particular lineage within the densely branching tree of life, and find ancestors who were fish, amphibians, reptiles, early mammals, early primates, apes – and then you. (And by the way, you're still an ape, just a very special one.)

I've written this book not only to help you reconnect with your very own origin as a human – from the point at which one of your mother's eggs was fertilised by one of your father's sperm – but also to let you reconnect with your ancestors. This book will take you on a tour of your body, starting with your head and moving all the way down to your toes and out to your fingertips. In the first few chapters, we'll focus on the earlier ancestors, like worms and fish, but we'll gradually get to more recent tiers of your family tree. By the time we get to talk about your extremities we'll be looking at the hands and feet of hominin ancestors and your closest living ape cousins.

The story of how a human body develops is (I hope to persuade you) the most fascinating narrative that science has to offer us. Each of us has been on this journey, from a single cell to a complex organism made up of hundreds of different types of cell, comprising something like 100 trillion cells in total. But each of us is also the product of evolution, and we're far from being a work of perfection, as we shall discover. Millions of years of evolution have produced something which works, certainly, but which is constrained by its history and the way it's constructed. The more I delve into the structure and the workings of the human body, the more I realise what a cobbled-together hodge-podge of bits and pieces this thing we each inhabit really is. It is brilliant, but it is also flawed. Our evolutionary history is woven into our embryological development and even our adult anatomy in surprising ways; many of our body's flaws can only be understood in an evolutionary context. Having said that, our ancestors left us gifts as well as glitches, and there are traces of extremely ancient ancestors in our bodies, in the shape of the developing embryo, and embedded in our DNA.

This is an exciting time in evolutionary biology, if exciting times mean lots of new questions. We're still trying to understand whether evolution proceeds gradually or takes leaps and bounds, and just how predictable evolution is. We're still exploring just how much the form and function of our bodies has been shaped by nature and nurture: how much it is constrained by its evolutionary past and the genetic programme which guides its development, and how much it has been influenced by environment and natural selection.

In telling this story of 'the making of us', through evolution and embryology, we'll explore our own anatomy and meet ancestors from our own evolutionary past as well as pioneering scientists who form the crew on this voyage of discovery. But the central character in the story really is YOU. It's about *your* evolutionary heritage, and it's about your own embryological development, when you grew and changed, parts of you folding like origami, until you were shaped like a human. It's the closest we ever come, as humans, to a transformation as profound as that from a caterpillar into a butterfly. Each of us underwent this transmogrification, from a single egg to a flat disc, to a hollow tube, to a little creature with stumpy arms and legs, to something that looked recognisably human – in the space of just two months after conception.

This is the *best* creation story, because it is true. It's also packed with quite bizarre revelations. In your DNA, there are traces of a common ancestor you shared with a fruit fly. At one point in your development, it looked like your embryo was about to grow gills. And the tools our ancestors began to make and use, millions of years ago, ended up *changing* their anatomy – helping to make your hands what they are today. This scientific story, pieced together from many different sources of evidence, is more extraordinary, more bizarre, more beautiful, than any creation myth we could have dreamt up.

A BRIEF HISTORY OF IDEAS

The origin of a new human, or in fact of any organism, was one of the great scientific mysteries until really quite recently. In the fourth century BC, Aristotle wrote *On the Generation of Animals* – the first scientific book about embryology. In it, he suggested that male semen activated female menstrual blood to create an embryo. Although that might seem like an odd proposition to us today, if you think about it, it's based on very reasonable assumptions; it assumes a link between sex and pregnancy – and that much, of course, is true. Long before anyone would look down a microscope and see a human egg, his idea about menstrual blood made a lot of sense because menstruation ceases when a woman becomes pregnant.

For many scientists over the ensuing centuries, the fact that there were no known specific precursors of an embryo, beyond bodily fluids, was not necessarily a problem. Some animals were even believed to originate from inanimate materials, so that flies could be generated spontaneously from putrid meat, for example. Aristotle's theory of development, which he called 'epigenesis', suggested that a complex human body could develop out of the mixing of simple fluids, of semen (and remember that he didn't know about sperm, semen was just a homogenous, milky fluid as far as he was concerned) and menstrual blood. His was a theory that would go largely unchallenged for two millennia.

The ancient Greek 'father of medicine', Hippocrates, had suggested that conception required both a male and a female seed, but Aristotle's idea – that male semen was the key ingredient in making a baby – became more influential. In the mid-seventeenth century, though, William Harvey obviously doubted this explanation, as he went about investigating the generation of animals through dissection, but although he was convinced that there must be a female 'egg', and even that it must originate in the ovary, he couldn't find it.

We all know how conception happens now, and it seems blindingly obvious. But the tale of the discovery of the origins of a human life is a fascinating one, and it depended on being able to *see* what was happening – on a minute scale. The discovery depended on technology which would enhance the optical abilities of the human eye, allowing it to see far smaller objects than it could ever manage without extra sets of lenses. Simple magnifying glasses were around from at least the sixteenth century, but it's not at all clear who invented the first microscope. Galileo is probably more famous for inventing the telescope, but he also invented something that he called an *occhiolino* (literally 'little eye' in Italian, but now meaning 'wink'). By the early seventeenth century, Galileo's 'little eye' had become known by the name we recognise: a microscope. Later that century, Robert Hooke used microscopy to study the hidden details of familiar objects – fleas, nettles and bee stings – publishing his findings in a beautiful book called *Micrographia*.

Meanwhile, on the other side of the North Sea, a Dutch draper called Antonie van Leeuwenhoek became obsessed with his hobby of making tiny glass lenses and using them to build microscopes. As he peered through the tiny lenses, he started to see all kinds of minute details and objects that no one had ever seen or recorded before. He saw *Volvox* algae, minute planktonic animals, flies laying eggs, human red blood cells and microscopic details of the spleen, muscle and bone.

And van Leeuwenhoek was also the first person to see a human sperm. Just imagine how amazing that would have been. It's hard, because *we* know that sperm exist, but forget that for a minute; it's 1677 and you're van Leeuwenhoek, and you're fascinated by the possibilities of the world of microscopy. You know that semen somehow helps to produce babies, so you get hold of some (I'll leave those details to your imagination) and you look at a small drop of that milky fluid using your microscope. You stare down, astounded by the sight that greets your eyes. The whole field of view is buzzing with movement. You can make out individual, tadpole-like cells, thrashing their tails furiously. They seem to be micro-organisms, like the protists you've already discovered (and written letters to the Royal Society about). But these 'animalcules' came from a human.

What's really astounding, though, with the benefit of hindsight, of course, is that neither Leeuwenhoek nor the scientists at the Royal Society in London, to whom he wrote describing his observations, immediately realised the importance of his observations: here was half of the secret of conception.

Another Dutchman wins the accolade of *almost* finding the human egg. He was a doctor called Regnier de Graaf, and in 1672 he published a treatise on female reproductive organs, including a description of the development of follicles inside the ovaries of rabbits. Those small balls of cells – present in human ovaries as well – would end up bearing his name: Graafian follicles. De Graaf also observed tiny spheres inside the Fallopian tubes after follicles had ruptured, and he deduced that the follicles – and the spheres – must contain eggs. But it wasn't until 1827 that someone would identify the mammalian egg itself.

That someone was Karl Ernst von Baer. As his name suggests, von Baer's ancestors were German, but he was born in Estonia, which, in 1792, was part of the Russian Empire. While a professor of zoology at Konigsberg University, von Baer studied embryology and in 1827 he discovered the mammalian ovum, nestled inside the Graafian follicle of the ovary.

Fantastic. These scientists seem to have cracked it: there's an egg, and there's a sperm. Together, they combine to make an embryo. Except, again, it's only that easy with our perspective from the present, with all that we now know. Perhaps Aristotelian ideas were so firmly entrenched that it was impossible to believe in an equal contribution from sperm and egg of whatever it was that would lead to the creation of a new individual. And so the scientific community became split into two camps: the ovists and the spermists. The ovists saw sperm merely as a force to 'awaken' the egg. The spermists saw the egg as just a source of nutrition for the new life created by the sperm.

The identification of sperm and egg also meant that Aristotle's idea of epigenesis, of complex new life somehow developing out of simple fluids, was pushed to one side. But this still left a similar conundrum to be solved. How could a complex organism develop from things as apparently simple as a sperm and an egg? For many scientists in the seventeenth and eighteenth centuries, the answer lay in the theory of preformationism. This theory suggested that the complexity already existed, in miniature, in the precursor of the embryo (in the egg or the sperm, depending on whether you were an ovist or a spermist). The most extreme version of this theory suggested that an entire, preformed person – a tiny 'homunculus' – was present inside the sperm. The Dutch lens-maker Nicholas Hartsoeker (who learnt his trade from Leeuwenhoek) drew such a homunculus, curled up tightly in the head of a sperm.

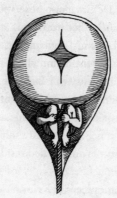

Homunculus – after Hartsoeker

The French philosopher and priest Nicolas Malebranche pushed the idea of preformationism even further. In 1674, he proposed the theory of *emboitement*, suggesting that every individual started off 'boxed up' inside their mother's egg. He wrote:

One sees in the germ of the bulb of a tulip the entire tulip. One also sees in the germ of a fresh egg . . . a chicken which is perhaps entirely formed.

He concluded that 'all the bodies of men and of animals which have been born . . . have perhaps been produced as long ago as the creation of the world'. In other words, everyone who has ever lived (and will ever live) was already there, in miniature form, packed up inside Eve's ovaries, like the most incredible set of Russian dolls. This theory of preformationism was also known back then as 'evolution', which was an apt term, as it means 'to unfurl' or 'to unroll'. Today, of course, that word means something very different. To someone who believed that the world was only a few thousand years old, perhaps *emboitement* seemed truly feasible. This was also before cell theory would set a lower limit of size, so it was possible to imagine that such tiny, preformed beings could exist.

But this is not to suggest that those early embryologists all believed in such an extreme form of preformationism. Anyone who had looked down a microscope and seen an early embryo would have known that it did not look like a tiny, preformed individual – at least not for the first few weeks of life. Although some of the more extreme aspects of preformationism seem ridiculous to us today, that lot were certainly onto something, because at the core of their argument was the idea that a complex organism could not appear from something completely unorganised, completely homogeneous. And of course they were right; it was just going to take a bit more time before anyone discovered the molecule that carried the information needed to form a completely new body.

THE BEGINNING OF YOU

We now have a much better understanding of what happens to create a new human. Your genetic identity, as a new individual, was sealed at the moment one of your father's sperm swam up into one of your mother's oviducts of Fallopian tubes to meet an egg which was making its own way down towards the uterus.

Imagine that egg: it has burst free from its home in the ovary, carrying with it a crowd of smaller cells. It has entered the funnel-like opening of the oviduct that is fringed with finger-like fimbriae, and now it is moving along, helped by the oviduct's lining of tiny, hair-like cilia which beat to create a current in the fluid inside the tube.

Now envisage one sperm, swimming hard, its tail thrashing furiously, up through the canal of the cervix, through the cavity of the uterus and into the oviduct (aka Fallopian tube). It's probably taken a few days to get this far. This sperm reaches the egg first through a mixture of luck and prowess; in a single ejaculation, a few hundred million sperm will have been projected into the vagina, but although they've already travelled a long way from their testicular home, they still have far to go. Many will perish before they even make it out of the vagina into the narrow corridor through the neck or cervix of the uterus. If it's the wrong time of the month, the sticky mucus in the cervix forms a barrier, preventing the sperm from progressing any further. Around the time of ovulation, though, the cervical mucus becomes more slippery and stringy. (This is an age-old way of predicting the days of optimum fertility within a woman's monthly cycle. Cervical mucus changes from being thick and sticky to being stretchy and egg-white-like in consistency. The German word for this property is 'spinnbarkeit', which means 'spinnability'.) As they travel through the cervix of the uterus, into the cavity within its body, more sperm will be left behind while others thrive in this environment and thrash their tails ever more wildly, swimming up, up and out into one of the oviducts. The egg sends out chemical signals to help the sperm choose the correct oviduct. The number of sperm that reach the egg itself is a tiny fraction of those that were ejaculated. Perhaps only one in a million sperm will make it this far. But the competition is still far from over.

Hundreds of sperm arrive at the egg at about the same time; the egg is surrounded by tiny sperm on all sides, but the egg only needs one of those sperm to make it through. Some will get through the cumulus oophorus – the halo of cells around the egg which have been left attached to it since the moment of ovulation when the egg burst free from the ovary. Once through those cells, the sperm arrive at the zona pellucida, a thick, gel-like layer surrounding the membrane of the egg. Now there's no escape; the zona pellucida traps the sperm. The heads of the sperm get stuck to the gel: sugary proteins in the gel attach to protein receptors in the membrane of the sperm, like tiny keys fitting into locks. And those keys really do unlock something: they trigger the release of enzymes from the tip of the sperm, allowing one sperm to penetrate right through the zona and reach the cell membrane of the egg itself. Now the membrane of the sperm is touching that of the egg. The membranes fuse, and the two cells – the tiny sperm and the huge egg – are united as one.

This was the moment when you were conceived: this extraordinary but everyday occurrence, hidden away in a dark recess of your mother's body. But it's also pertinent to remember that this fertilised egg is *not* a person. It is just a cell. There's no guarantee, at that point in time, that this cell will develop into a whole organism. It's only with hindsight that you can say: this is where I started.

At the moment of contact, when the membranes of the sperm and egg fuse, three things happen. The sperm continues swimming into the egg, leaving its membrane behind: a mere shell, abandoned on the surface of the egg. Inside the egg, and lying close to its enveloping membrane, are tiny bags of chemicals. Now these tiny bags fuse with the egg's membrane, emptying their contents into the space just underneath the zona pellucida and transforming the zona, making it set hard, to prevent any more sperm reaching and fusing with the egg. This is important: all the egg needs is one set of chromosomes to complement its own – 23 chromosomes to pair up with the 23 it already has. If any more sperm were to fertilise the egg, this would mean the arrival of extra sets of chromosomes, and the chances of developing into a viable embryo would be ruined.

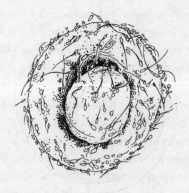

Sperm arriving at the egg

Lastly, as the sperm enters the egg, the maternal DNA undergoes a final stage of preparation before the maternal chromosomes can pair up with the male set.

As the sperm swims into the egg, its tail falls away and degenerates. On a minute scale, this is like launching a satellite into orbit. When the rocket reaches its destination, the launch vehicle separates from its payload – the satellite itself. Inside the egg, the payload is the head of the sperm, which contains a package of chromosomes and is the essential information needed by the egg if it is to become an embryo. This package of genetic material starts to swell as the chromosomes within it are unpacked. The double strand of DNA which forms each chromosome is then unzipped – and this is where the magic of DNA comes in. Tiny DNA building blocks (or nucleotides) attach to each half of the 'zip' until two new zips are formed. A similar doubling-up is going on inside the package of chromosomes belonging to the egg. Then both packages – one from the egg, one from the sperm, each bulging with a double set of chromosomes – push together and fuse. Thus the DNA from egg and sperm – from mother and father – is brought together for the first time.

There are now 46 pairs of chromosomes, already enough DNA for two cells. The double chromosomes line up in the middle of the egg, as

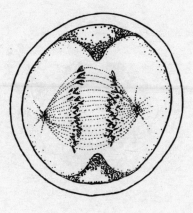

Pairs of chromosomes pulling apart on the spindle, as the egg starts to divide

if pairing up to dance. They assemble on a scaffold known as a spindle, made of incredibly thin tubes of protein. Then each double chromosome splits in two and the two chromosomes pull apart from each other, moving to opposite ends of the spindle-scaffold. At the same time, the membrane of the fertilised egg, where all this is taking place, is pinching in, until it becomes dumbbell shaped. Eventually it pinches off completely, cleaving in the middle, and two cells are born. It's about 24 hours since the sperm first fused with the egg. At the end of your first day of life, you are two cells.

But there's no resting now, because each cell is working hard. DNA is duplicated again and the cells divide. Three days later, you're a ball of sixteen cells. You're described perfectly and poetically by the technical term for what you've become: a morula, from the Latin for 'little mulberry'. And while all this has been happening, you've been on the move. Swept along by waves of contractions in the muscle wall of the oviduct, and by tiny waving hairs called cilia on the inside of the Fallopian tube, you've almost reached the cavity of the uterus. You may be just a ball of cells at this point, but already the cells both on the outside and inside of the ball

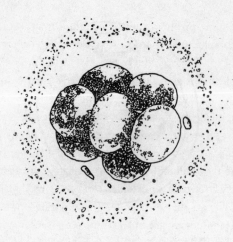

Morula

are destined for different fates. Your outer cells will help to form the placenta: your life-support mechanism while you're in the womb; your inner cells will form the real you: the embryo.

Just a week after fertilisation, as you drift into the uterus, a fluid-filled space appears inside the ball of cells. You're no longer a morula; you're now a blastocyst – a word from Ancient Greek which means 'hollow bud'. The inner mass of cells isn't spread evenly around the inside of the outer cell mass, it's clumped at one end of the blastocyst. You've developed polarity. It may not seem like an important development, but this unequal distribution of cells means that the developing embryo now has an orientation. Regardless of which way up the blastocyst is floating in the fluid inside the oviduct, it now has its own *internal* orientation. The cells of the inner cell mass which face the new cavity will have a different fate from those towards the edge.

It's time for your short journey of perhaps ten centimetres to end. You come to rest against the inner lining of the uterus and almost immediately your outer cells begin to invade that lining, which marks the beginning of the formation of a placenta.

There are so many opportunities for things to go wrong in embryology. The more complex an organism is, the more chances there are for its development to go awry. But in fact, right from the start of development, even before it starts to get particularly complicated, there are ample opportunities for fatal mistakes, and the resting place of the blastocyst is one of them. Your blastocyst may have settled safely, but it doesn't always turn out that way. Blastocysts can get stuck in the oviduct, or, much more rarely, go the wrong way and end up in the mother's body cavity, where they will still attempt to implant, wherever they lodge. This is what's known as an ectopic (from the Greek for 'out-placed') pregnancy. Whereas the uterus is designed to enlarge with the growing embryo, other tissues are not so accommodating. An ectopic pregnancy can be a very dangerous error: a growing embryo stuck inside the oviduct is likely to cause blood vessels to rupture, and this can lead to

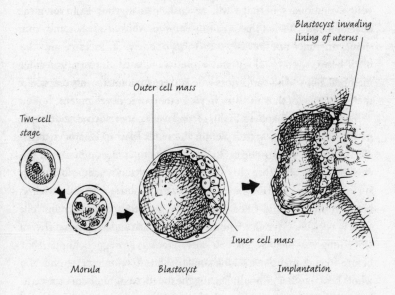

The first week of development: from a two-cell embryo, to a morula, to a blastocyst, to implantation in the uterus

disastrous internal bleeding. In the absence of surgical intervention, such a haemorrhage is likely to be fatal.

With the benefit of hindsight, we know that the blastocyst-you will have implanted safely in your mother's uterine wall. As you enter the second week of development, the cells of the inner cell mass (which are still proliferating) resolve themselves into two layers: an upper layer (epiblast) and an under layer (hypoblast). This differentiation depends on the position of cells relative to the blastocyst cavity. Cells from the hypoblast spread out to line the inside of the original blastocyst cavity. From now on, this space is known as the yolk sac, even though it contains no yolk, because you're a placental mammal with no need for an extra bag of nutrients. But right here, we're discovering something really important about how embryological development makes a body. No organism is 'designed' by evolution from scratch. It's all about tweaking and tinkering with what's already there. This means that some aspects of your evolutionary heritage will be enshrined in your embryological development. The fact that, as a tiny embryo, you *had* a yolk sac, even a small, un-yolky one, reveals something about your ancestry and the links between embryology and evolution. You might be a placental mammal, but you had ancestors who laid eggs, complete with yolks, and there's an echo of that ancestry in your embryonic development.

Something else happens in this second week of development: a brand-new space begins to appear within the outer layer of blastocyst cells. This space is the amniotic cavity. To begin with, it faces one side of the developing embryo, but this is the fluid-filled sac which will end up surrounding you, and in which you'll float until you are born. Sandwiched between the two fluid-filled spaces – the amniotic cavity and the yolk sac – is a double-layered, roundish disc of epiblast and hypoblast. There's still a long way to go before you start to look even vaguely human, but before you get there, you'll look similar to the developing embryos of a whole host of other animals. During the fourth week after conception, a human embryo looks very much like a fish embryo at a similar stage of development. In the fifth week, when your limbs begin to sprout, you are pretty much indistinguishable from a chick embryo; a couple of weeks later, you still look very much like any other mammal embryo. You could almost be mistaken for a pig, dog or mouse embryo, although your head shape and the five digits on each hand and foot reveal that you're a primate.

So, it's not just the yolk sac that harks back to our evolutionary past. The embryo's passing resemblances to ancient ancestors and other animals alive today led to one of the most ignominious theories in the history of embryology: recapitulation.

REMEMBERING PAST LIVES

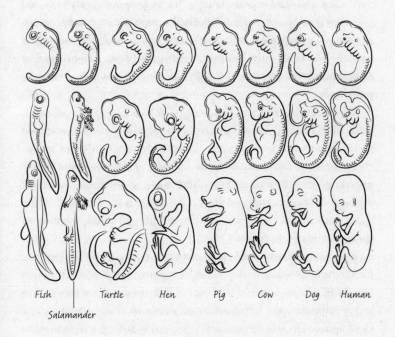

Fish | Turtle | Hen | Pig | Cow | Dog | Human

Salamander

A drawing taken from Haeckel's illustration showing similar stages of embryonic development in eight different animals

Looking at tiny embryos of different animals, it's impossible to ignore the fact that there are some 'echoes' of evolution to be seen in the embryo. In a human embryo, there's a stage when it has what look remarkably like the primordia of fish gills, and a stage at which its heart looks a lot like an embryonic fish heart. So is it really possible that human embryos are 'remembering' or 'recapitulating' their evolutionary past?

Working and writing in Germany, in the latter half of the nineteenth century, Ernst Haeckel was a pioneering biologist who discovered and named many new species, and he also threw himself into promoting Darwin's ideas about evolution. He's remembered, too, as the man who invented the theory of recapitulation – and got it so wrong – although he wasn't the first scientist to come up with this idea.

Aristotle classified organisms according to degrees of complexity and perfection, and he believed that humans developed through similar stages as embryos, eventually reaching perfection – as a human, of course. But Aristotle was really only drawing an analogy between various stages of embryonic development and his classification of animals, ranked by complexity.

This linear classification of animals – all animals – was a hugely influential idea. The preformationists (who imagined a whole person curled up inside a sperm or an egg) saw the entire history of life on Earth as encapsulated at the moment of the creation of the world and simply unfurling since that time. Everything was pre-ordained, and every organism was connected in a great chain of life, a *scala naturae*. There was a sequence of increasing complexity and perfection, reaching its apogee, of course, in civilised Man.

In the late eighteenth and early nineteenth centuries, the biologists of the German school of *Naturphilosophie* rejected the idea that evolution proceeded in a pre-ordained way, but still believed that it was moving in a very specific direction: towards increasing complexity and consciousness. Once again, Man was the pinnacle. It was one of these *Naturphilosophen* who came up with the idea of recapitulation: the physician and embryologist, Johann Friedrich Meckel. His name is enshrined in two embryonic structures which all medical students learn about: Meckel's cartilage (in the developing mandible) and Meckel's diverticulum (a small out-pouching of the intestine). Meckel saw a deep connection between the great chain of life and the way in which an apparently simple embryo became more complex during its development. For Meckel, embryos really were experiencing a re-run of evolution, sped up and on a very small scale.

Not all of the *Naturphilosophen* were wedded to this idea of recapitulation. Karl Ernst von Baer, the discoverer of the mammalian ovum, studied chick embryos and spotted a few major problems with recapitulation: firstly, embryos never looked completely like the adults of any other animals. Secondly, structures didn't necessarily appear 'in order' in the embryo, following the same sequence as the *scala naturae*. Thirdly, and most importantly, von Baer saw that embryonic development was actually about something very simple turning into something much more complex. As even 'primitive' animals like fish are actually very complex, recapitulation just didn't make sense. But in his rejection of recapitulation, von Baer had hit upon a real, fundamental law of biological development: differentiation. In his chick embryos, he had seen with his own eyes how a complex organism developed from simple beginnings.

All of these developmental theories in the first half of the nineteenth century were constructed on a theoretical framework that was just about to be taken apart. That framework was biblical creationism, which carried with it the dogma of the immutability of species. What this meant was that the appearance of common structures in embryology and adult animals was a mark of a divine plan.

In 1859 Charles Darwin published *On the Origin of Species by Means of Natural Selection*, and the entire rich tapestry of biology was re-hung on a new frame. In fact, the central thesis of the *Origin of Species* had been launched on the world in the previous year, when a joint paper by Darwin and Alfred Russel Wallace had been presented to the Linnean Society of London. Rather oddly, with hindsight, this failed to make much of a splash, but the *Origin* made sure that the idea of evolution through natural selection was noticed. Earlier ideas about a sequence of immutable species on a *scala naturae* were now replaced with *real* sequences of species evolving over time in a branching tree of life (although the idea of linearity would die hard, and still clings on to this day).

In the *Origin of Species* Darwin wrote about the striking similarities between embryos of different animals, illustrating his point with an anecdote about the famous anatomist, Louis Agassiz, who 'having forgotten to [label] the embryo of some vertebrate animal, he cannot now tell whether it be that of a mammal, bird, or reptile'. Darwin realised that the resemblances between embryos could provide important clues about evolutionary relationships between animals – clues that later became obscured by the appearance of specific adaptions in adult animals. In a creationist view of biology, the similarities between embryos (and adults) had represented an abstract connection between animals in the mind of a creator. Under the new evolutionary paradigm, those resemblances spoke of real, physical links between ancestors and descendants.

In Germany, Ernst Haeckel was a great proponent of Darwin's theory of evolution and wrote his own popular books on biology, morphology and evolution. Despite von Baer's objections, the theory of recapitulation was still going strong in the mid-nineteenth century, and Haeckel gave it an evolutionary bent. Haeckel believed that evolutionary change occurred through new modifications being added on at the end of embryological development. This meant that the embryological development of an organism would reflect the exact sequence of its evolutionary development. So a human embryo, for instance, might be expected to pass through stages when it looked like a fish, an amphibian and a reptile before it started to look more like a mammal. Haeckel summed up this idea as 'ontogeny recapitulates phylogeny' – in other words, embryological development recaps evolutionary history.

Haeckel's recapitulation theory, known as his Biogenetic Law, became incredibly popular, and many biologists were won over by the resonance between ontogeny and phylogeny, explained so neatly by recapitulation. But the theory was due for a spectacular fall from grace. Around the turn of the twentieth century, the rise of experimental embryology and the appearance of a new science of inheritance called 'genetics' spelled the end for Haeckel's theory. Embryologists started to examine the mechanics of development, moving around pieces of early amphibian embryos and watching what happened as a result. Genetics showed that new changes were not just tacked onto the end of embryological development: genes are present right from the moment of conception, and mutations could affect development at any point. The crucial ideas behind recapitulation, that extra features could only be added *at the end* of embryonic development, and that embryos passed through stages equivalent to *adult* ancestors, no longer stood up to scrutiny.

The dramatic collapse of recapitulation makes the subject a delicate one. Mentioning its name is almost like uttering some biological profanity. It's now such a thoroughly disreputable theory that its only use seems to be as a cautionary tale. But there *are* parallels between embryological development and evolutionary history. Haeckel was wrong; animals don't have the equivalent of their adult ancestors telescoped into their embryos. But von Baer (who had been largely forgotten about in all the furore), and Darwin too, had been right. Resemblances between embryos were there because of common ancestry.

Darwin and Alfred Russel Wallace both came up with the theory of evolution by natural selection by looking, in the main, at the variation in anatomy and physiology – and embryological development – of living animals. This is incredibly important, and something that is often overlooked by creationists, because it means that the theory does not depend on the fossil record nor on any scientific advances since the Victorian period. The most elegant explanation for the patterns we see in animals living today is that they are all related: they are all twigs on a huge, evolutionary tree of life. In the latter half of the nineteenth century it was clear to biologists and geologists that extinct animals, known in the form of fossils, were also part of this great tree. But since Darwin and Wallace's paper in 1858 there have been many wonderful discoveries of extinct fossil organisms which provide links between groups of animals. We now have fossils of fish with limb-like fins, like *Tiktaalik*, as well as fossils of early amphibians like *Acanthostega*, which show us what the first limbs looked like. We have feathered dinosaurs which tell us about the origin of birds. We have fossil ancestors of whales, which still possess legs. We have fossils of reptiles which look like ancestors of mammals. We currently have knowledge of around twenty fossil hominin species: a six-million-year-old family tree of two-legged apes which includes our own ancestors.

As well as this bounty of fossil evidence, we can now examine the structure of any body in much greater detail than the Victorians could have dreamt of, using techniques like electron microscopy and immunohistochemistry, staining cells according to specific proteins they produce. And of course there have been great leaps forward in understanding the nature of inherited characteristics, with the discovery of DNA, the elucidation of the function of genes (a major area of research which still continues), and the reading of entire genomes (again, something which is really, if not embryonic, certainly in its infancy).

Embryology itself has been reinvigorated by advances in histology and genetics. Experiments in the second half of the twentieth century began to reveal how cells 'decided' what tissues they would develop into, or where they would end up in the body. The study of embryology was transformed after the discovery of DNA as the 'code of life' – now it wasn't just about describing *how* an embryo formed over time, but finding out what genes drove the process. While von Baer was able to peer down his microscope and find resemblances in the early embryos of what would become very different animals – chicks, fish, humans – DNA sequencing uncovered a deeper vein of similarity, written into the genetic codes of animals.

The modern science of embryology reveals how the genetic code of an organism is translated into proteins which build a body. In order to reconstruct the tree of life, we can now use not only comparative anatomy but go deeper, using comparative embryology and comparative genomics. Family trees of species built from DNA sequences offer more insights into evolutionary history than those based on just anatomy. This new blending together of embryology, genetics and evolution (which has become known as 'Evo-Devo') has the power to answer important questions about embryological development and the evolutionary history of organisms. Today, a new generation of embryologists, whilst firmly rejecting Haeckel's Biogenetic Law, is busy discovering even more links between ontogeny and phylogeny, between embryology and evolution.

It is through these similarities with other animals, in our adult anatomy, in our embryological development, in our genetic code, that we are able to understand our own place within the great tree of life – the *arbor naturae*. We are just a twig on that tree, rather than evolution's ultimate destination (it doesn't have one). Looking carefully at your anatomy, it will also become clear that you're certainly not the 'pinnacle of evolution' that you might like to imagine you are. You're far from a perfect creation, more like a rag-bag collection of bits and pieces, the result of millions of years of tinkering. But as far as natural selection is concerned, you'll do, and that's why you're here today.

Embryology and evolution explain why your body is the way it is. The structure and function of your adult body is a product of your own embryological development and your evolutionary past. From head to toe, you are the living embodiment of that history.

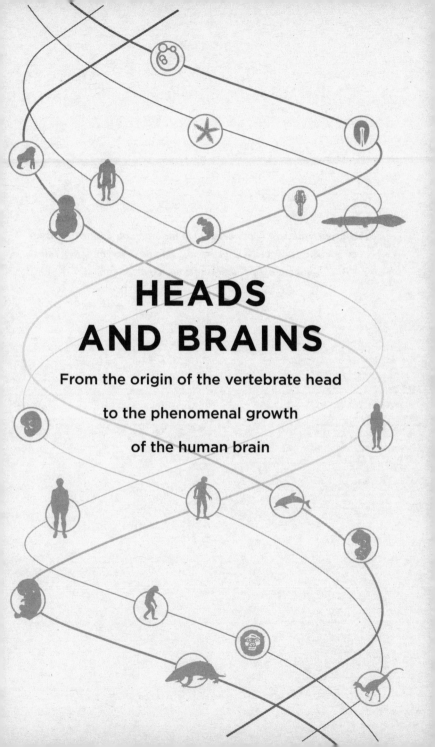

HEADS
AND BRAINS

From the origin of the vertebrate head

to the phenomenal growth

of the human brain

'People often ask how I got interested in the brain; my rhetorical answer is: "How can anyone NOT be interested in it?" Everything you call "human nature" and consciousness arises from it'

VILAYANUR S. RAMACHANDRAN

THE FIRST HEAD

t's a strange question, perhaps, but have you ever considered *why* you've got a head? It's obviously not a uniquely human characteristic; most of the animals we're familiar with have heads. Certainly, having a head seems to be a prerequisite if you're any sort of vertebrate: a fish, amphibian, reptile, bird or mammal, that is. Lots of invertebrates have heads too, but some of them don't. In order to answer the question 'why do we have heads?', it would be useful to know *when* our ancestors first developed this bit of anatomy.

I thought that I had been provided with an answer to this question in biology classes at school. I learned that vertebrates had evolved from much simpler organisms, similar to living sea squirts, which are close cousins of vertebrates. There is probably a zoologist somewhere who is terribly excited by sea squirts, but, especially in comparison with vertebrates, I don't think it's unfair to say that these animals are sedentary and boring. These almost inanimate creatures lurk on the bottom of the sea, with no intent, stuck to the ocean floor, sucking in seawater and filtering out small particles of food from it.

I got up close and personal with a sea squirt, in the interests of delving into some comparative anatomy, while filming with the BBC. I was just off the coast of a small, almost deserted island called Mnemba, east of Zanzibar; the sea was azure, incredibly clear, and we were there with snorkels and an underwater television camera. I jumped out of our boat into the ridiculously blue sea, with a mask and snorkel, and put my head under water. I had never snorkelled on a coral reef before and the sight that met my eyes left me gasping. I looked up and called out to the crew on the boat, 'There are *hundreds* of fish down there!' I think everyone else had experienced this sort of thing before, at least, they were far too nonchalant to catch my enthusiasm, but I wasn't letting that dampen my first-time wonder.

Diving down about four metres to the reef itself, I searched for a sea squirt and eventually found one, bringing it ashore in a bucket. Examining the soft creature, about the size of a small potato (and only marginally more interesting), I found two openings in its rubbery flesh. These are somewhat delicately referred to as siphons, but in any other animal they'd be called the mouth and anus. Between these two openings, sea squirts have a very simple, U-shaped gut. The squirt sucks seawater in through its mouth (its buccal siphon), and particles of plankton get trapped in the mucus which lines its gut, while the rest of the water passes through to be expelled out of the other (atrial) siphon. A sea squirt doesn't have any organs of special sense such as eyes, but it does sense the world around it in more subtle ways, with receptors scattered across the surface of its body that respond to light, touch and various chemicals. The life of a sea squirt does seem remarkably dull: it's effectively just a gut. It does not crawl, swim, see or think. It just sits

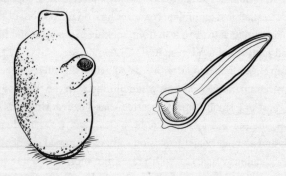

An adult sea squirt (left) and its tadpole larva (not to scale)

there, getting nutrition from seawater in a manner which can barely even be described as 'eating'.

But there's a phase in the sea squirt's life when things are, briefly, more interesting. As a 'tadpole larva', the sea squirt *swims*. It's a tiny, fish-like thing, swimming around, moving of its own volition. This larva has both a tail and a head, but it will soon lose both. When it grows up and settles down, just three days later, it becomes headless. In the mid-nineteenth century, a Russian embryologist called Alexander Kowalevsky (who was a student of Haeckel) noticed that the sea squirt larva was rather special: it had a strengthening rod of tissue (called a notochord) and a nerve tube in its tail. These features were already known to exist in vertebrate embryos, so although the adult sea squirt lost its notochord and nerve tube, its larva revealed it to be very closely related indeed to vertebrates. It looked like Kowalevsky had stumbled on a biological secret: it seemed that vertebrates might have evolved from a neotenous sea-squirt-like ancestor – a Peter Pan sea squirt that forgot to grow up

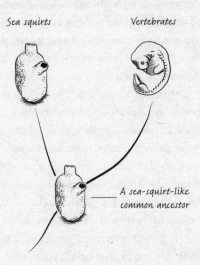

Sea squirts   Vertebrates

_____ A sea-squirt-like
          common ancestor

A family tree (with earliest ancestor at the bottom) showing how vertebrates could have evolved from a sea-squirt-like ancestor

and settle down, and instead remained a freely swimming 'larva' for its whole life. This was still the story written in textbooks in the 1980s, when I was studying biology in school.

Recent research has cast doubt on this sequence of events. It's always an occupational hazard in science: what seems like a great theory today, based on available evidence, may be consigned to the dustbin of history tomorrow, when a new piece of evidence awkwardly refuses to be shoehorned into the existing paradigm. So, new research has suggested that, although ascidians like sea squirts are certainly related to us, they're not very useful if we want to know what that ancestor of vertebrates looked like. Studies of both sea-squirt DNA and morphology (body shape) have revealed that these animals have actually become *simplified* over time. The common ancestor of ascidians and vertebrates was probably not a sedentary creature like the adult sea squirt (and that adult form must have evolved later), but an animal that swam freely throughout its entire life.

In fact, there are some animals which are closely related to both vertebrates and sea squirts, and spend their lives swimming: they're called lancelets. They are small creatures which look remarkably like fish, but they are not fish – they don't have spines (or skulls, for that matter). They are called lancelets (because they are pointed like lances), and are also known as *amphioxus* (from the Greek for 'double-pointed'). Together, vertebrates, sea squirts and lancelets are classified as 'chordates' because they all possess a notochord. There are a few other defining characteristics of chordates, including a hollow nerve tube, gill slits and a tail. We should pause here for a moment's reflection, because you may be thinking: 'Hang on a minute! I'm a vertebrate, which means I'm also a chordate. But I don't have gill slits and a tail! And I don't even think I've got a hollow nerve tube or a notochord, either.' The key to this conundrum is that, to be a chordate, you just had to have those features *at some point in your life*, and in this respect, you're a bit like a sea squirt, which possesses these characteristics as a larva but loses them as it grows. As an embryo, you did indeed have a notochord (which forms a

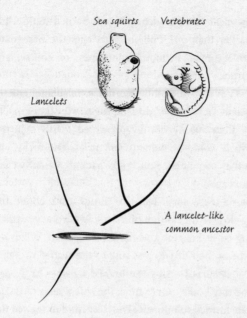

A family tree (with earliest ancestor at the bottom) showing the evolution of vertebrates and sea squirts from a free-swimming, lancelet-like ancestor

sort of 'pre-spine'), a hollow nerve tube, a tail and gill slits. You *are* a chordate.

Asserting our own chordate credentials means that we can see how we relate to all the other animals on this planet – both living and extinct. Chordates are just one of around 35 large groups, or 'phyla', within the animal kingdom. The classification of vertebrates, sea squirts and lancelets as chordates was proposed by none other than Ernst Haeckel, back in 1874. The origin of this group (our own phylum) extends far back in time to the geological period called the Cambrian, which started 542 million years ago. Before this time, life on Earth consisted almost entirely of single-celled organisms, but this period saw a huge proliferation and diversification of multicellular life, mostly living in the sea, in what has become known as the 'Cambrian explosion'. Although there's now clear fossil evidence of soft-bodied multicellular animals existing earlier than 542 million years ago, the first fossils of many major animal phyla, including chordates, of course, are found in Cambrian rocks.

At the very end of the twentieth century, palaeontologists prospecting among ancient rocks in the Yunnan province in southern China discovered more than 300 beautifully preserved fossils of a tiny, fish-like animal, which would be named *Haikouella*. The rocks, and therefore the fossils they contained, dated right back to the early Cambrian, 530 million years ago.

So imagine these small fish-like things, each about an inch long, swimming close to the bottom of a shallow sea and occasionally resting and settling on it. Imagine a whole shoal of them. When oxygen levels dip, they die in their droves and sink to the seabed, which is such fine mud that it is stirred up by the merest whisper of a current, gently burying the tiny bodies. Over time, the soft tissues of those not-quite-fish are transformed to mineral. The silky mud of the sea floor hardens into a fine-grained mudstone, preserving the anatomy of these animals in exquisite detail.

Even though each tiny fossil creature measures only a few centimetres long, it's possible to discern the stripes of individual muscle blocks, a strengthening rod along its back called a notochord, the gut, frilly arches which would have supported gills, and even a minute brain. The mouth is fringed with twelve short tentacles. One fossil preserves something which looks intriguingly like an eye. It's an odd little thing which looks a little like a worm, a little like a fish. In fact, it's neither. What it looks most like is a lancelet. The anatomy of living lancelets is incredibly similar to the ghostly anatomy preserved in those 530-million-year-old Chinese rocks. *Haikouella* is the earliest example of a chordate that has been discovered, and it most definitely has a head. Whereas today's lancelets are our distant cousins, *Haikouella* – or at least, something very much like it – may have been an ancient ancestor from which humans (and all other chordates) evolved. This means that

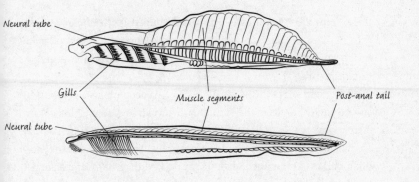

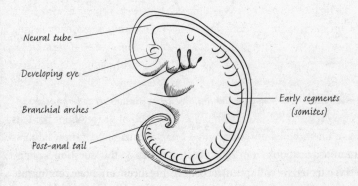

Chordate characteristics in (from top to bottom) Haikouella, a modern-day lancelet, and a human embryo towards the end of the third week of development

heads (or at least our chordate version of heads) have been around for at least 530 million years.

For the origins of this free-swimming lifestyle – which seems to be linked to possessing a head – we have to look wider and deeper into the evolutionary tree of life, and even further back in time. Swimming down into these murky depths, it can be difficult to see the answers. Textbooks are cautious about the origin of chordates and vertebrates and the picture has changed considerably in just the last decade or so. I have an

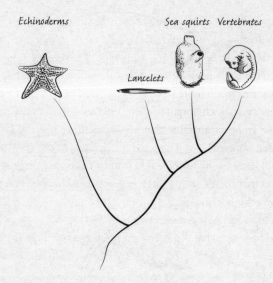

Echinoderms *Sea squirts* *Vertebrates* *Lancelets*

'Martian' cousins – a family tree showing how vertebrates turn out to be quite closely related to echinoderms

edition of a textbook from 2001 which embarks on this question, saying 'In this chapter, we will speculate concerning the invertebrate origins [of vertebrates]'.

Of all the other groups of animals alive on Earth today, among the most closely related to us chordates are the spiny-skinned echinoderms, including sea urchins, starfish, brittle stars, sea lilies and sea cucumbers. Although variations have evolved, all of these animals are based originally on a 'pentaradial' body plan: a five-fold symmetry. It's hard to imagine something less like a human, indeed less like a fish or even a lancelet, than a starfish. Richard Dawkins went as far as to call them 'Martians'. They are very odd indeed compared with vertebrates, and even compared with other invertebrates too, which also tend to be bilaterally symmetrical, with a front and back, left and right side, in contrast to the weird radial symmetry of starfish. But there is a point at which starfish do resemble chordates – and that's when they are tiny embryos.

The early embryonic development of starfish is similar to that of chordates (including us), and starfish larvae – rather like those of sea squirts – are free-swimming creatures. Unlike their adult selves, starfish larvae *are* bilaterally symmetrical, with a head end and a tail end. This seems much more familiar. So, instead of evolving from a sea-squirt-like larva which 'forgot to grow up', could the earliest chordates have come from an ancestor that was something like the larva of a modern echinoderm?

For zoologists trying to solve this riddle of chordate origins, the big problem is that there's such a large gap between chordates on the one hand and echinoderms on the other. There are enough anatomical, embryological and now genetic clues to show that these two groups are closely related, but what did their common ancestor look like? Was it more like an echinoderm or a chordate? Did starfish evolve from free-swimming ancestors, or did we evolve from something more sedentary, a sessile filter-feeder whose larvae forgot to grow up? Until quite recently, the idea of a 'Peter Pan' larva (this time, of something echinoderm-like rather than sea-squirt-like) was popular. But this is where genetics really has come to the rescue, because even if two animals look superficially quite different, you may be able to find genetic similarities running deep which speak of an evolutionary affinity: a common ancestor.

There's another piece in this puzzle: a branch of the family tree that I have, somewhat disingenuously, omitted to mention until now. Those spiny-skinned sea urchins, starfish and sea lilies have some close relations called hemichordates ('half-chordates') or 'acorn worms'. These solitary worms live in mud or under rocks in shallow waters, where they grow up to 2 metres in length, and they have a weird mix of chordate and invertebrate characteristics. Acorn worms apparently got their name from the shape of their proboscis, which looks more like a cartoon rocket to me. The whole worm, it has to be said, resembles nothing less than a gigantic sperm. (I've drawn a sperm next to it, just to drive the point home. There really is no point to this, beyond a superficial resemblance, but if I'd been in charge of naming acorn worms, they would

An acorn worm (on the left) and a human sperm (on the right, not to scale)

have been called sperm worms, which is obviously far superior because it rhymes.)

Ten years after Ernst Haeckel founded the phylum Chordata, in 1884, the English biologist William Bateson (famous for coining the term 'genetics') added the acorn worms to it. His rationale for doing this was that he thought he could see definitive chordate features in acorn worms. These weird creatures had stacks of gill slits like a lancelet, and they had a nerve cord along their backs which looked like it could be hollow in places. Bateson also believed that a short rod of tissue projecting into the acorn worm's proboscis could be its notochord. Later biologists questioned this interpretation, and by the 1940s acorn worms had been kicked out of the chordate clan. Acorn worms themselves were blissfully unaware of their demotion. They continued living in their muddy burrows and minding their own businesses, and it seemed that humans had all but lost interest in them.

But in the last couple of decades, acorn worms have really come back on trend. The modern science which Bateson named has unlocked the secrets hidden in their DNA. Acorn worms are still not chordates, but as slightly more distant relatives they are even more useful, as they contain clues to the origin of our phylum.

The position of acorn worms in the family tree finally became clear when their DNA was sequenced: they are the sister group of the echinoderms. Looking at differences in the string of 'letters' (the four

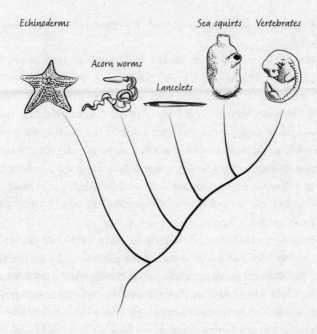

Echinoderms   Acorn worms   Sea squirts   Vertebrates   Lancelets

A family tree (with earliest ancestor at the bottom) showing how acorn worms are closely related to chordates – the common ancestor of both was probably a worm

nucleotide bases) that make up DNA, biologists can work out the relationships between different animals, drawing up 'family trees' which are part of the great evolutionary tree of life. The differences in DNA reflect actual evolutionary relationships between animals. For instance, a vertebrate will be genetically more similar to a sea squirt than either of them is to an acorn worm because the vertebrate and the sea squirt share a much more recent common ancestor. Comparing DNA between animals might seem like a completely new science, but while it is nothing short of revolutionary, it is really an extension of what evolutionary biologists have always done – comparing and contrasting bits of different animals. It's just that, instead of body parts, it is genes and base-pairs in the DNA sequence which are being compared. It's comparative – molecular – anatomy.

DNA sequences help us to build family trees, but genetics can also shed light on whether a particular structure in one animal really is equivalent to – or 'homologous with' – a similar structure in another animal. Homology exists in different animals because of shared ancestry. Darwin recognised, for instance, that a human hand, a bat's wing and a porpoise's fin are homologous structures, all inherited from a common ancestor. But sometimes homologies can be hard to spot, and the best way of testing whether structures really are homologous is by looking at which genes are 'turned on' (in the jargon: 'expressed') when those structures are being formed in the embryo.

Similar genes are expressed on the inner surface of the developing gill slits in chordates and acorn worms: these gill slits really are homologous. But different genes are expressed in the chordate notochord and the structure which Bateson thought was the equivalent structure in acorn worms – they are not homologous. It also turns out that he was wrong about acorn worms possessing a hollow nerve cord; what they have instead is a diffuse epidermal 'nerve net' which lies just under the surface of their bodies. It was just a little part of this nerve net that Bateson mistook for a hollow nerve cord.

By putting all the clues together, from comparative anatomy, embryology, genetics and palaeontology, it seems that at last we have an answer to the question of when and why our ancestors first developed heads. The proboscis of our distant relative, the acorn worm, with its mouth tucked in just behind, barely qualifies as a head, but it seems that chordates (and echinoderms), rather than evolving from the larva of an altogether more sedentary adult sea creature, probably descended from a worm that learned to swim. (And, rather interestingly, this means that the radial symmetry of adult starfish was something which must have evolved later – they come from bilaterally symmetrical ancestors just as we do. We've hung on to 'primitive' bilateral symmetry while starfish came up with an eye-catching new design). The development of a head is linked to the development of swimming in our ancient ancestors. Having a head isn't just about being mobile, because sea urchins and starfish certainly move. (There was a particularly impressive time-lapse sequence of sea stars in the BBC series *Life*, which is still available to view on YouTube.) But starfish – as shown beautifully in that time-lapse sequence – can move in any direction they like. Could you look at a starfish and honestly say which was its front end? The key to the question 'Should this animal have a head?' turns out to be: 'Does this animal have a front?'. And the faster you move, the more head-like your front is likely to become. For a free-swimming animal, it helps to have your senses stacked up-front, in a head, where you first encounter novelty in your environment. Of course, it also helps if you have a brain, to process all that information coming in from your head-mounted sense organs.

ANCIENT AND EMBRYONIC BRAINS

In those 530-million-year-old rocks from the Yunnan province we see the earliest evidence of a chordate, the earliest evidence of a real head, and the earliest evidence of a brain. Like any respectable chordate, *Haikouella* has a hollow neural tube, and the front end of this tube is slightly thickened and divided into three segments. It's not much to look at, perhaps, but this thick front end of the neural tube is *Haikouella*'s brain. Amazingly, your own brain – as large and complex as it is – started out in the same way: as a thickening of an embryonic neural tube.

Earlier in this book, we left you as a developing embryo, implanted in the wall of your mother's womb. The inner cell mass of the mulberry-like morula had become a flat, two-layered disc, sandwiched between the yolk sac and the newly formed amniotic cavity. The upper layer of this disc is the epiblast, and the lower layer is the hypoblast.

Now, as you embark on the third week of your embryonic development, some very interesting events are about to take place. It's time for

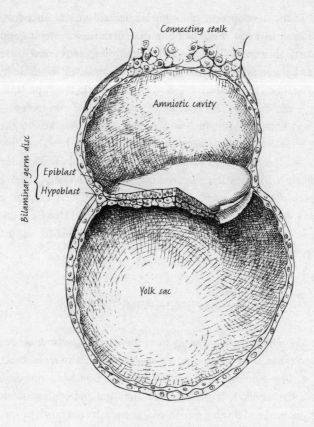

Connecting stalk

Amniotic cavity

Bilaminar germ disc

{ *Epiblast*

Hypoblast

Yolk sac

In the second week of development, the embryo is a two-layered disc, sandwiched between the amniotic cavity and the yolk sac (which comes from the original blastocyst cavity)

an out-of-body experience. Imagine your embryo scaled up massively, and you are floating in the amniotic cavity, looking down at the surface of the epiblast, which is now shaped more like a 2D pear than a round disc. (And remember, this peculiar, alien-like flat object is *you* barely two weeks after conception.) This flattened pear shape already has a front (which is the wider part) and a back, a left and a right. Something strange is happening to the surface of the epiblast: you can see it buckling into a groove which runs down the midline of the embryo, like a geological fault. Epiblast cells are multiplying and moving towards this fault line and then disappearing down through it. They will push the

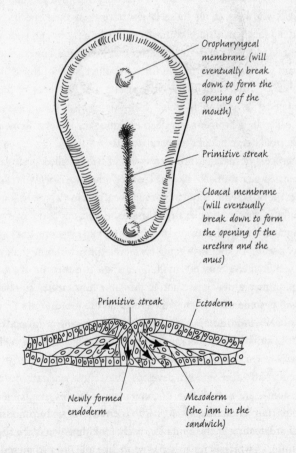

Oropharyngeal membrane (will eventually break down to form the opening of the mouth)

Primitive streak

Cloacal membrane (will eventually break down to form the opening of the urethra and the anus)

Primitive streak   Ectoderm

Newly formed endoderm

Mesoderm (the jam in the sandwich)

The surface of the epiblast showing the primitive streak (above) and a cross-section through the primitive streak (below) to show cells piling in to the inside of the germ disc, to make the new mesoderm and endoderm

hypoblast cells out of the way, forming a new layer in its place. They will also push out to form a new middle layer, sandwiched between the original epiblast and the new 'under layer'.

The result of all this moving and multiplying is that the embryo's original two layers have been converted into three: ectoderm, mesoderm and endoderm. You are now a three-layered or trilaminar germ disc. Cells within each of these layers have their own specific fates mapped out for them. Endoderm will end up lining your gut, your lungs and your bladder. Mesoderm will eventually become bones, muscles and blood vessels. Ectoderm will form the outer layer of the skin and create nerves. And right there we're seeing another facet of human embryological development which connects us with distantly related animals, like an echo from our deep evolutionary history. Remember the epidermal nerve net of the acorn worm?

This process of making three layers out of two is called 'gastrulation', because in simpler animals – like the lancelet – this movement of cells also produces the earliest gut. It's an important event because it lays the foundations for building the complex organism that is you, and the way you did it also signifies a fundamental division in the animal kingdom, marking an ancestral parting-of-the-ways going back even further in time than we've already ventured, to almost 600 million years ago. In one group of animals, gastrulation makes new layers and at the same time creates an opening which will become the mouth – these animals are protostomes ('mouth first' in Greek). Protostomes are a huge group, including arthropods (the massive phylum which includes insects, crustaceans and spiders), molluscs and several worm phyla. In the other group of animals, the deuterostomes ('mouth second'), the mouth develops later, as a separate opening. Deuterostomes are a small clan in comparison, and this group includes our own phylum, chordates, together with acorn worms and echinoderms.

For chordates like us, gastrulation is also the time when the first of our defining characteristics appears. As the middle layer of mesoderm forms (like jam being injected into the middle of a sandwich), some of it thickens up to form a rod along the central axis of the embryo. This rod is the notochord (Greek for 'back-cord') and it's crucial to the formation of yet another chordate feature: the neural tube. It's almost as though the notochord is 'talking' to the upper layer of ectoderm, lying above it, telling it to start changing. And in a way, that's exactly what's happening, it's just that the conversation is mediated by chemical signals. DNA contains a set of instructions for building an embryo, and this is how the pattern is generated: when particular genes are switched on, cells will produce signalling proteins which tell other cells what to do next.

In this case, as chemical signals flow out from the notochord, the overlying ectoderm starts to thicken up. Soon, a spoon-shaped area of ectoderm stands proud of the rest of the ectoderm layer. This spoon shape then starts to crease in the midline, forming a long groove flanked by two ridges. The crests of the ridges begin to curl inwards and eventually meet in the middle, sealing off the groove and forming a tube.

In the 1960s, embryologists predicted that the formation of structures like the neural tube – indeed, all of embryonic development – must depend upon the existence of signalling proteins. But it wasn't until the 1990s that the molecules themselves, and the genes which encode

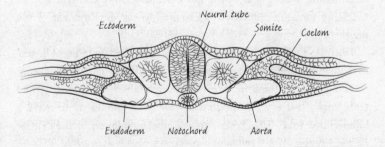

A cross-section through a human embryo towards the end of the third week of development. The embryo is still a flat disc at this stage, but the neural tube has formed and the mesoderm on either side of the neural tube is condensing to form bumps called somites. Blood vessels – an aorta on each side – are already forming. The coelom is the precursor of the body's cavities: the membrane-lined pockets around the lungs, heart and intestines

them, began to be elucidated. In a normal embryo cells in the upper (dorsal) part of the neural tube will become sensory neurons, whereas cells in the lower (ventral) neural tube are destined to become motor neurons. This pattern persists in the mature spinal cord, where sensory neurons (bringing information in) are grouped towards the back and motor neurons (carrying signals out to muscles) are grouped nearer the front of the cord. Experiments in chick embryos showed that, if the notochord was removed, the ventral motor neurons failed to develop. It seemed likely that the notochord was producing a chemical signal which affected cells differently along a concentration gradient. Eventually the gene and then the signal itself was tracked down, and given the wonderful name of 'sonic hedgehog' (in honour of the blue-haired video game character).

Although you only possess a notochord for a brief period of intra-uterine life, and as a respectable vertebrate you've built a much more substantial vertebral column in its place (of which more later), you still needed that notochord early on. Without it, you wouldn't have developed a spinal cord or a brain. Think of the scaffolding used to build a tower block: it will be taken down and removed completely when the building is complete, but it was essential during construction. In the same way, the notochord is essential to the development of the neural tube.

The tube doesn't seal up all along its length at the same time: it begins to close in the middle of the embryonic disc – this is actually the future neck region of the embryo. Then it zips up, in both directions, and its ends finally seal over, in the fourth week of development, to make a blind-ended tube. This hollow tube is the foundation of the central nervous system: your brain and spinal cord.

If the neural folds fail to meet and fuse, the result is a 'neural tube defect'. In Britain, about eight in 10,000 babies are affected. The range and impact of these defects is very wide. If the front end of the neural tube fails to close, the embryonic brain cannot develop. This is a defect known as anencephaly (from the Greek for 'brainless') and babies with

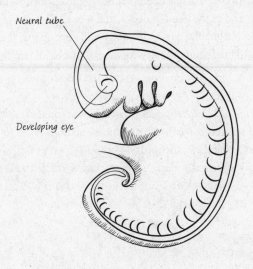

In the fourth week of development, the neural tube closes over at the head and tail end, and the embryo is no longer a flat disc: its sides have curled down until they meet in the middle, so that the outer surface of the embryo is now entirely ectoderm

this problem die at or very soon after birth. Severe defects like this are usually picked up in prenatal ultrasound scans. Failure of closure in the rest of the neural tube is known as spina bifida (literally: 'split spine'), and this can range from a severe defect where the legs are paralysed, to a hidden cleft in the vertebrae which never causes any problems.

As a four-week-old embryo, your brand-new neural tube, slightly wider and thicker at the front end, bears a striking resemblance to the neural tube of primitive chordates like the living lancelets and the very ancient *Haikouella*. Your brain still has a huge amount of developing left to do, but the lancelet brain is only ever a slight expansion of the neural tube, and barely more complex than the spinal cord behind it. In vertebrates, including us of course, a more complex brain starts to form in the embryo: the anterior end of the blind-ended neural tube swells to form a series of three connected bubbles or vesicles which will develop into the forebrain, midbrain and hindbrain. The development of the hindbrain is controlled

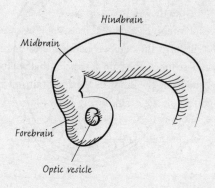

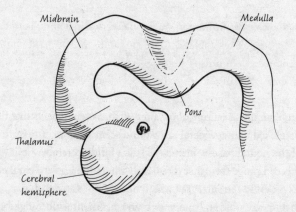

The developing embryonic brain, late in the fourth week of development (top) and in the sixth week (bottom) when the forebrain has developed into the cerebral hemispheres and the thalamus, and the hindbrain into the pons and medulla

by a set of pattern-generating *Hox* genes with a very ancient pedigree – versions of these genes are also present in the fruit fly, with which we share a common ancestor going back some 800 million years. Areas of the forebrain and midbrain are specified by genes which have arisen more recently in our evolutionary history, after the split between protostomes and deuterostomes, but before the origin of chordates.

As the front of the forebrain continues to swell it also separates into two lobes, which will form the cerebral hemispheres or, simply, 'cerebrum'. Just behind these lobes, the pair of vesicles that will become your eyes are growing out from the sides of the forebrain. In many vertebrates, the early pattern of the brain during embryonic development is still easily recognisable in the adult brain. At first glance, the human brain seems to be an entirely different beast, but this is because the forebrain has become blown out of all proportion. Our massively over-inflated cerebral hemispheres overlap the rest of the brain, but if you look under the brain, or cut it open, you can see the structures that have developed out of those early embryonic vesicles. While the developing embryonic human brain looks very much like that of a shark, it's only a passing resemblance. The human brain ends up being much more complex than a fish's brain. We come from a lineage of animals which has specialised in expanding their cerebral hemispheres. Mammals have relatively larger cerebral hemispheres than reptiles. Placental mammals (the group which includes us, along with most other mammals) have larger cerebral hemispheres than mammals such as duck-billed platypuses; primates have large cerebral hemispheres compared with most other mammals, and humans have taken this expansion to an extreme.

In fish, the brainstem (midbrain plus hindbrain) is the biggest part of the brain. The forebrain is small, with relatively large olfactory lobes. The cerebrum is slightly enlarged in amphibians compared with fish: it has to deal with more sensory information as well as controlling more elaborate musculature in limbs. The cerebral hemispheres get even bigger in reptiles (and birds), bulging out to the sides and covering up

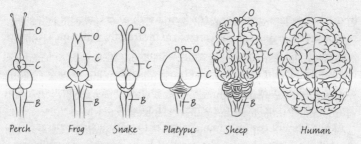

Perch   Frog   Snake   Platypus   Sheep   Human

O—Olfactory tract & bulb
C—Cerebral hemisphere
B—Brainstem

A selection of brains (not to scale) showing the relative size of the cerebral hemispheres; in humans, looking at the top of brain as shown here, all that can be seen are the hugely inflated cerebral hemispheres

the underlying thalamus. The cerebrum of mammals – particularly placental mammals – gets crazily big. In addition to this increase in size there's another important change in mammals: the growth of a brand-new outer layer of brain tissue – or 'neocortex'. The neocortex of mammals has grown so big that it overlaps the more evolutionarily ancient parts of the brain. You have to look hard for those archaic areas of cortex, but they are there. Tucked away on the underside the brain, on the inner edge of the temporal lobe, is the olfactory cortex, receiving that most ancient of senses, smell, and also the hippocampus, which is involved in memory.

The neocortex of mammals contains neuron cell bodies that have been pushed to the outside of the brain. The neocortex has three major roles: it receives and makes sense of sensory information coming in from the whole body; it sends out motor signals to muscles; and it collates sensory information and sends it off to be stored as memories. Under the microscope, the neocortex is seen to have six layers, and in most mammals it is highly folded. This folding increases the volume of cortex that can be fitted inside the skull – imagine crumpling up a sheet of paper and rolling it into a ball. In fact, the area of the human cortex would be equivalent to a very large sheet of paper – about 2400cm^2, which is slightly smaller than A2 size (or four pieces of A4). There are an estimated 86 billion neurons in a human brain, with 13 billion neurons in the human cerebral cortex. We don't have the biggest brains of any animal, though, as the brains of elephants and whales, for example, are much bigger, but we do have extremely large brains relative to our body size.

MAPPING THE HUMAN BRAIN

I have two very precious and rare objects which I was lucky to have made for me while I was in the process of making films about human anatomy and evolution with the BBC. One is a reconstruction of my skull, and the other a reconstruction of my brain. Both are based on data from a detailed MRI (magnetic resonance imaging) scan taken of my head, and have been made into solid objects using 3D printing.

I'm usually quite sanguine about seeing bits of my internal anatomy. Again, mostly in the course of filming or teaching, rather than for any medical reason, I've probably seen more of the structure of my own body than most people have. Through ultrasound, I've seen my heart pumping and the muscles and nerves in my neck; using a tiny pill-cam I've watched a video of my intestines, filmed from the inside; and thanks to MRI, I've looked at my uterus and ovaries, my head and my larynx. But when those head MRI scans were used to reconstruct my skull, I was quite freaked out by the result. Unpacking the finished alabaster-white object from its box, I came face to face with myself. I'm used to looking at skulls, real skulls, but this model really affected me. Like the cadaver tombs of the fifteenth century, designed more to remind the living of their mortality than to honour the dead, and the figurines used in the Mexican Day of the Dead, my 3D printed skull was a potent reminder of death – an extremely personal and powerful *memento mori*.

In less than a hundred years from now, once the worms have done a thorough job, the real me would look just like it.

The other 3D printed part of me, my brain, I find less arresting, but still utterly intriguing. I have it sitting on the desk in front of me as I write. How strange to think that all these thoughts are taking place as nerve impulses race along and leap across synapses inside its real counterpart, residing safely inside my skull.

My 3D brain model is a good reconstruction. Particularly on its upper surface, I can clearly see the gyri and sulci – which are, respectively, the folds and the grooves between them – which characterise the walnut-like appearance of the human brain. I can even see the central groove or sulcus, which makes its way up each side of the brain, from the bottom to the top, all the way up to the great cleft between the two cerebral hemispheres. The wriggle of grey matter in front of this sulcus is the precentral gyrus, which contains the cell bodies of motor neurons which send their axons down into the brainstem and spinal cord, synapsing there with secondary motor neurons which reach out with long fibres to all the skeletal muscles in my body, even down to the ones which move my toes. Behind the central sulcus is the postcentral gyrus; this receives incoming, sensory signals from right across the surface of the body. The arrangement of neurons in these gyri, with respect to the body parts they reach out to, or receive information from, is far from random. The connections with the rest of the body can be mapped onto the cortex here and are often drawn as a homunculus (that word again, although this time it represents a real pattern of connections in the brain, rather than an imagined 'little man' curled up in the head of a sperm).

The central sulcus also forms the division between two of the lobes of each cerebral hemisphere: the frontal and parietal lobes. There are two remaining lobes on each side: the temporal lobe (which is a wedge lying under the parietal and frontal lobes, under cover of the temporal bone), and the occipital lobe at the back (also lying underneath the bone of the same name).

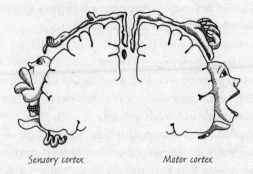

Sensory cortex   Motor cortex

These homunculi show how the body is mapped out in the postcentral gyrus (primary sensory cortex) and precentral gyrus (primary motor cortex): highly innervated areas appear larger

It's quite incredible to think that the real version of this organ, the lump of living nervous tissue inside my skull, is behind everything I'm aware of, and doing, right now. Think about what *you're* doing at this moment: the parts of the frontal cortex which control your fingers are active as you turn pages or scroll down. As well as controlling voluntary movement, your frontal lobes are also involved with concentration, memory, recognition and emotions. Your parietal lobes are busy processing stimuli, including visual ones – at the moment, taking in what's on the page or screen in front of you. Your temporal lobes are crucial for language, allowing you to recognise words in both the written and spoken form. Your occipital lobes are receiving information from your retinae and forming an image: although you may *feel* as though you are seeing with your eyes, you're really only converting light into electrical signals in your eyeballs; it's at the very back of your brain that those signals start to make sense, where they are made into a picture.

There's also partitioning of function between the right and left cerebral hemispheres. A crossing-over of neurons in the spinal cord means that the right side of the brain controls movement in the left side, and vice versa. For other functions, the division is asymmetrical: the 'creative' right side (in most people) is involved in spatial perception, art and music. The left side is involved with more 'rational' jobs like controlling language and marshalling logical thought. But the two sides are also in constant communication with each other. In fact, all the time there are messages being sent from one area of cortex to another, and not just across that central divide. While the cortex is part of the grey matter of the brain, stuffed full of nerve cell bodies, the messages travel in the white matter, which is made up of the 'wires' of the brain, the long processes of nerve cells called axons. Each axon is covered in a layer of insulation, just like the plastic around electrical wires, except that around nerve fibres it's made of a particular fat called myelin. The bundles of myelinated fibres forming connections between areas of the cortex, between the two cerebral hemispheres, and between the cortex and deep islands of grey matter, make up the white matter of the brain. There's still very little myelin in a newborn baby's brain; much of the growth of a baby's brain is due to nerve fibres becoming insulated with myelin, rather than new nerve cells appearing.

Looking again at that 3D reconstruction of my brain, at the back of it, I can see my cerebellum nestled under my occipital lobes. This is where sensory information from the body is put together with commands to muscles, issuing forth from the cerebral hemispheres, in order to coordinate movement and maintain balance. This is why people with damage to the cerebellum find it difficult to move in a smooth and precise manner. However, functional scans of human brains performed while people are doing various tasks show that the cerebellum has a cognitive as well as a purely motor function. If I then turn my 3D brain upside down, I can see the parts of my brainstem: the midbrain, pons and medulla. I can even see, just about, where my evolutionarily ancient olfactory cortex and hippocampus lie.

Our current understanding of what happens where in the brain – what various 'modules' are responsible for – has been informed by the development of technologies which allow us to visualise the living brain, such as functional MRI (fMRI). But the first clues that different regions of the brain were responsible for different functions came long before the invention of brain scans, from case studies of people who had suffered localised brain damage, such as a stroke or an injury. The most famous of these early documented cases is that of the American railway worker, Phineas Cage. On 13 September 1848, 25-year-old Gage was injured in an explosion which drove an iron rod right through his skull. Miraculously, he survived, and he was examined by a physician called John Harlow, whose report on Gage's accident was published in 1868, in the *Publications of the Massachusetts Medical Society*.

Harlow describes how the iron bar had penetrated Gage's head:

> The missile entered by its pointed end, the left side of the face, immediately anterior to the angle of the lower jaw, and passing obliquely upwards, and obliquely backwards, emerged in the median line, at the back part of the frontal bone, near the coronal suture.

Harlow goes on to recount how Gage quickly recovered consciousness and was able to speak within minutes of the accident, and how, having been transported in a cart to his hotel, he then got up and walked upstairs to his room. Harlow examined Gage's wounds: a hole in his cheek where the iron rod had entered, and a large oblong hole, measuring two by three and half inches, in the top of his head. Harlow felt inside the wounds. He could push his entire right index finger down into the hole on top, and his entire left index finger up into the cavity in Gage's cheek, but could not quite bring the tips of the two fingers together. Harlow cleaned up the wounds.

Over the ensuing months, Harlow checked up on Gage. In November of the same year, Harlow recorded that Gage 'continued to improve steadily,' and that he was in good physical health. Gage was apparently keen to return to work, but the railway company would not take him on again.

Harlow recorded that before his injury Gage had 'possessed a well-balanced mind and was looked upon by those who knew him as a shrewd, smart business man, very energetic and persistent in executing all his plans of operation'. His bosses had 'regarded him as the most efficient and capable foreman in their employ'. But the damage to Gage's brain had altered his personality. Now, he was 'fitful, irreverent, indulging at times in the grossest profanity . . . impatient . . ., at times pertinaciously obstinate, yet capricious and vacillating . . . his mind was radically changed, so decidedly that his friends and acquaintances said he was "no longer Gage"'.

Phineas Gage spent the next four years travelling around New York and New England, spending some time with Barnum's Circus, taking his tamping iron and his story with him. From there, he travelled to Chile and then, eight years later, to San Francisco, where his mother and sister lived. He drifted from one job to another, unable to stick at anything for long. In 1861, he began to suffer fits, and in May of that year, twelve years after his railroad injury, following a series of severe convulsions, he died.

Although an autopsy was not performed on Gage's body, his mother let Harlow open Gage's grave and retrieve his skull, 'for the benefit of science'. The passage of the tamping iron was clear to see: up through the left maxilla, through the back of the left eye socket and out through the frontal bone at the top of the skull. The bar would certainly have passed through the left side of the frontal lobe of Gage's brain, and Harlow speculated that it would have also injured the left temporal lobe, and possibly ripped through into the lateral ventricle – the fluid-filled space inside the temporal lobe.

More than 150 years after Phineas Gage suffered and survived his extraordinary injury, a team of anatomists and radiologists from Harvard Medical School re-examined Gage's skull, this time subjecting it to a computed tomography (CT) scan. Using the scan data to make a virtual 3D reconstruction of Gage's skull, the team could then model the trajectory of the 3cm-thick tamping iron. It became clear that the temporal lobe of Gage's brain had in fact been spared because the iron bar only passed through his left frontal lobe. This trajectory would not have directly affected the motor area of the frontal lobe, which explains why Gage did not suffer paralysis as a result. The rod had closely missed both the internal carotid artery at the base of the skull and the superior sagittal sinus (a vein enclosed in layers of meninges) at the top of the cranium. If it *had* damaged either of these large blood vessels, this would have led to massive bleeding, and Gage's injury would undoubtedly have proved fatal.

We have come a long way since Phineas Gage and his tamping iron revealed the frontal lobe of the brain to be somehow involved with reasoning, decision-making and social behaviour. Much more recent functional studies of the brain have shown that these 'higher' cognitive functions take place in the anterior part of the frontal lobe, called the 'prefrontal cortex', just in front of the motor cortex. At the time, scientists didn't take much notice of the lesson of Gage's injury: that there was functional specialisation in the brain, whereby different areas are responsible for different functions. It seems incredible that the importance of Harlow's findings was missed, especially as his insight seems like common sense to us today.

Before the nineteenth century the brain was largely assumed to be a homogenous blob in terms of function, much like the liver. Phrenologists had suggested that different areas of the brain related to different cognitive functions – but their reading of bumps in skulls was shakily grounded in pseudoscience.

The first widely recognised evidence of functional specialisation in the brain came from the research of the French doctor, anatomist and anthropologist, Paul Broca. By examining the brains of patients who had suffered brain damage and had problems with speech, Broca identified an area of the frontal lobe which seemed to be responsible for spoken language. In 1861 he published his findings in a paper entitled *Sur le principe des localisations cérébrales* ('On the principles of cerebral localisation') – coincidentally, this was the same year that Phineas Gage died.

In 1874, a German neurologist called Carl Wernicke suggested that the upper part of the temporal lobe was responsible for the understanding of language, again based on deficits in patients who had suffered damage to that area of the brain. Although modern neuroscientific studies have refined our understanding of the relationship between anatomy and function in the brain, the regions identified by those pioneers are still known as 'Broca's area' and 'Wernicke's area'.

Like the map of a new country, gradually emerging from surveys and exploration, the cortex was mapped. Eventually, the brain, like a globe, had all its continents, countries and islands drawn and defined. You can open any neuroanatomy textbook and see these areas, precisely circumscribed and neatly labelled. Except it's not that simple. Of course it's not, this is the *brain* we're talking about. There are some 13 billion neurons just in the neocortex, and so with an average of something like 7000 junctions or synapses per neuron, that makes for nearly 100 trillion synaptic connections in the cortex. These sorts of numbers make me nervous. Mapping the general areas of the brain related to various functions is one thing; trying to draw a real map of all those connections is another entirely. And yet some neuroscientists are brave enough to try.

In 2007, a team of researchers at Harvard Medical School, led by Professors Jeff Lichtman and Joshua Sanes, invented a new imaging technique: a multi-coloured neuron mapping system designed to illuminate the complex connectivity of the brain. Using genetic engineering, they inserted extra genes into mouse genomes. These genes were instructions for producing different colours of fluorescent protein, which would make the mouse neurons glow. Random combinations of colours produced more than a hundred different hues which could be used to identify particular neurons. Looking at slices of mouse brain under the microscope, the colours helped researchers to accurately trace the long fibres of nerve cells, and to identify cells making synapses with each other. In a further stroke of brilliance, they christened the method 'Brainbow'.

A couple of years ago, I was lucky enough to meet Jeff Lichtman at Harvard. We chatted about the uniqueness of the human brain. We have such large brains, with so many neurons inside them, but Jeff was clear that what made us humans unique went beyond sheer brain size, or indeed sheer numbers of neurons: the important thing was the circuitry itself, the connectivity of the neurons. In primitive animals, at least primitive in this respect, like the much-studied worm *Caenorhabditis elegans* (a favourite of embryologists everywhere), with just 300 neurons in its brain, the connections were largely genetically determined. In primates, and in humans in particular, genetic programming set up a range of possibilities which were then honed through interaction of an individual with its environment. At birth, a human baby already has most of its neurons and far too many connections. This isn't a mistake – far from it – it's an essential part of how complex nervous systems develop. Superfluous connections get pruned back as the brain develops, based on experience. The result would be that each neuron would limit, but also strengthen, the connections it made with other neurons. It wasn't just about losing connections, but about re-deploying them in a more limited or strategic way. Jeff had seen something similar happening in the innervation of muscle fibres. At birth, each motor neuron innervated many muscle fibres and each muscle fibre was innervated by a range of different neurons. A competition would ensue, with each axon competing to contact the muscle fibre, but only one would win. (It's fascinating to think about development in this way: you're a complex colony of cells, and those cells have battled for survival.)

With the longest childhood of any mammal, humans have extended the time during which connections in the brain are whittled down, and in fact, we continue to do this – to learn – throughout life. Our brains, far from being genetically determined, are a product of nature *and* nurture. Genes set the range of possibilities, while interaction with the environment (including, of course, very importantly, culture) sculpts the connections in the adult brain.

For Jeff, the incredible complexity of the human brain also left room for free will. 'There are so many factors, so much complexity, that it's impossible to predict outcomes – either free will exists, or the complexity is such that it might as well exist'.

I asked Jeff about the task of mapping the human brain, and the peculiarity of using a network of neurons to image and try to understand a network of neurons. 'It's curiously circular,' he agreed. 'The idea that we're using this very complex machine to look at the complex machine that we're trying to understand. Especially when we're looking at the visual connections in the thalamus. We're using those connections in our own brains to interpret images of the connections . . .'

Jeff was obviously prepared to be daunted and mindboggled. He was also philosophical about the undertaking to deconstruct the human brain, to take it apart and understand its parts, and reassemble it. 'I think our minds are this physical thing. Although on top of that structure what we're uncovering is the dynamic process – the function of the nerve cells, the electric impulses that travel through these connections. But even as a neuroscientist, I find it troubling that this is all there is to the human brain, these connections.'

However we image the brain, and try to understand it, whether that's by creating a solid, 3D model of a brain from an MRI scan or mapping the neurons and their connections, it all seems to fall short of capturing the real essence of what's there inside our skulls. One day we may know about all those connections, and understand exactly how our brains function, in incredible detail, but I'm not sure even that will feel any different. It's as though the mind is determined to remain mysterious, experiential, unknowable.

Jeff Lichtman's work is essentially anatomy – at an electron-microscopic scale. But what do all those billions of connections actually do? One way of studying that is to go back to the origins of this science and look at people with brain damage and functional deficits, like

Phineas Gage. But now we also have the technology to see activity in living brains. It sounds gruesome, but there are now a few completely non-invasive ways in which neuroscientists can virtually get inside people's heads.

One way of looking at the function of the brain, non-invasively, is by recording its electrical activity, using electrodes stuck to the scalp: this is called electroencephalography (EEG). But it's difficult to precisely locate where in the brain the electrical activity is occurring. Other functional imaging methods measure brain activity using proxies. Positron emission tomography (PET) scans use the rate of glucose metabolism or blood flow as a marker of the metabolic activity in brain tissue. Functional magnetic resonance imaging (fMRI) can be used to identify active regions of brain tissue due to increased blood flow to those areas. Functional brain scans have revealed that, although it's possible to map function onto areas of cortex, the edges of such areas are indistinct, and what seems like one function may use several different regions of cortex. What's important is the connectivity between the areas.

There's still a huge gulf between Jeff Lichtman's fine neuroanatomy, where individual neurons, axons, dendrites and synapses are surveyed, and the broader brush-stroke pictures emerging from functional imaging. But the hope has to be that, at some point in the future, these two spheres of investigation will move closer together and we'll gain a high-resolution, detailed understanding of the brain's functional anatomy.

On the couple of occasions on which I've had fMRI scans, I've been amazed to see this visualisation of activity in my own brain. In the first scan, a few years ago, I was intrigued to see the areas of my motor cortex that control muscles in my hands lighting up when I moved my fingers. But this was what I expected to happen: I knew where the motor cortex was, after all. But the second scan, performed in 2011, was more surprising. I was about to find out something about the brain that has only been recently discovered: a secret that I certainly hadn't been let into when I was introduced to neuroanatomy at medical school in the early 1990s.

MIRROR NEURONS

On a chilly morning in January 2011, I took the train to London and arrived at Birkbeck College, with a mounting feeling of trepidation. There wasn't a clinical reason to have this scan, but there's always the worry, when you have such a test performed, that something unexpected and perhaps even sinister, might be discovered. Although, rationally, I'd rather know if something was wrong and I know that the chances of finding some pathology in this way are very slim, it's still nerve-wracking.

I was just walking up to the College entrance when I bumped into fellow medic-turned-television-presenter, Michael Mosley. He was just about to undergo a procedure – also for filming rather than any clinical need – which would involve temporary brain damage: disabling Broca's area so that he would not be able to speak. (A producer once asked if I would consider having this done, and I refused. Apart from sounding like an unnecessarily scary thing to subject yourself to in the name of television science, I couldn't help but be concerned about its effects – I know the risks are diminishingly low, but *what if* you didn't recover completely?) I told Michael that he was much braver than I had been. I didn't see him again that day, but I saw him many months later, and I am glad to say he appeared to have regained his powers of communication in full.

Once inside the building I descended to the basement where, tucked away at the end of a corridor, Dr Geoff Bird was waiting for me in the MRI suite. Geoff's research focuses on using fMRI to investigate how we copy others, but he also looks at where seemingly more nebulous cognitive functions – attention, emotions and empathy – happen in the brain.

I changed into clothes with no metal bits in them and removed all my jewellery, and then I was ready to be placed inside the enormously powerful magnet. I sat on the edge of the shelf-like bed which would carry me into the scanner while Geoff ran through the procedure for the test. Then I inserted some earplugs into my ears (MRI machines make very loud clanking noises) and lay back on the bed. Geoff placed a mask-like cage over my head, which contained a radiofrequency coil which was integral to the scanning process. With radiofrequency waves penetrating my head, the protocol reassuringly had built-in rest periods between scans so that my head didn't heat up too much. Now I really had gone past the point of no return, and Geoff left the room, shutting the heavy door behind him. But I was still able to communicate with him – and the film crew – in the control room, via the intercom.

Now that I was on my own in the scanner room, the machine ground into action, levitating me upwards then propelling me along until my head was well inside the giant doughnut of the scanner. A small, angled mirror positioned on the cage above my head allowed me look out of the scanner – I could just make out dim figures in the control room. Geoff had also hung an opaque screen at the mouth of the scanner, onto which my instructions were projected. We got started, and a series of words appeared on the screen: 'point', 'fist', 'stretch' and 'thumb', each encouraging me to make an appropriate movement with my right hand, returning it to a relaxed, resting posture between each posture. ('Point' and 'fist' are self-explanatory; 'stretch' meant splay your fingers out, while 'thumb' was a 'thumbs-up' gesture.) The words on the screen rattled along, and I barely kept up. Sometimes I'd try to anticipate the next word, and fail dismally, pointing instead of sticking my thumb up, splaying my fingers instead of balling them into a fist. I found that I had to concentrate really hard, but still I made mistakes, and I was swearing at myself inside the scanner. I hoped I wouldn't ruin the experiment. Each set of instructions and hand movements was followed by a short and less-than-entertaining video which just showed different hands making similar gestures. My role here was to stay still and watch. After a few cycles, I found even that to be quite difficult, as I got bored easily and my mind wandered off to think about other more interesting things, like lunch. Then I'd catch myself and make myself pay attention.

Later on, Geoff showed me the results of my scan in visual form: the active areas of my brain highlighted in red on a grey brain image. The results were arresting: there were areas in the frontal lobe, in the 'prefrontal cortex' which is involved with planning movement, which lit up when I moved my hand, and, remarkably, the same regions lit up again when I *watched* other people making the same hand movements on the video.

So here, in my own brain, was evidence of neurons which fired not only when I carried out particular actions, but also when I watched someone else performing the same actions. You've got those neurons, too. Everyone has. These are 'mirror neurons' and their existence makes cognitive neuroscience even more complex – and more exciting. In fact, you don't even have to *see* someone doing something for a mirror neuron to be switched on: they can also fire when the sound of a particular action is heard. This has been demonstrated with nut-cracking in monkeys, and in humans listening to speech. And these neurons are very clever – they are activated not just by simple actions, but by actions that are linked by a common goal. They seem to help us understand the intentionality of a particular action.

Just think about all the things you've ever learned by watching or copying someone else. In fact, that's far too much to think about, and it starts earlier than you can remember. Humans are exceptionally good at copying each other, so mirror neurons are extremely important (although not unique) to us, underlying our ability to imitate and learn from others.

The neuroscientist Vilayanur S. Ramachandran has argued that mirror neurons give us an innate ability for imitation, so that a newborn baby is able to copy his mother by poking out his tongue. This example grabbed my attention: I've poked my tongue out at both my babies, when they were very small and new, and I've been amazed by their ability to copy me. Just think about what this ability means: a baby doesn't need to practise in front of a mirror in order to replicate the action of poking her tongue out (in fact, she wouldn't recognise the person in the mirror as herself anyway). The baby's brain seems to be set up ready to make that link between the faces of others and her own.

Mirror neurons are activated when we watch or hear other people performing various actions or speaking particular words, but their function seems to extend much further. When we see someone hurt or upset, we empathise: we don't just understand their pain in an intellectual manner, we *feel* it. Ramachandran calls the mirror neurons 'Gandhi neurons': they blur the boundary between self and others. Humans have an extraordinary ability to work out what others are doing and thinking – to 'mind read' – another capacity which might depend on mirror neurons. So it seems that these special neurons could be essential to how we operate as a social species: able to empathise, cooperate and, of course, deceive each other. Ramachandran has also suggested that mirror neurons are crucial to human evolution much more generally, allowing the development of language, the flowering of culture and the rise of civilization.

I quizzed Geoff about mirror neurons. He was in the middle of researching the proposed link between mirror neurons and empathy, as well as the more controversial suggestion that autism could be linked to a lack of mirror neuron activity. It sounded logical in a way, but Geoff was yet to be convinced that there was in fact any convincing evidence for this connection; previous studies had produced mixed results. He was also sceptical about whether mirror neurons, and their particular functionality, were an innate feature of our brains. He thought that this could be a learned component of brain function – that positive feedback from copying others, for example, could encourage the development of such neurons. Was I encouraging my baby to copy what I was doing by smiling at him when he did so?

Mirror neurons are not unique to humans, as they exist in other apes, and in monkeys, too. But where did they come from? It's possible that they are, in themselves, an adaptation that has been favoured by natural selection. In this view, mirror neurons may be innate, and any individuals (monkeys or humans) possessing them and therefore an inbuilt ability to understand the actions of others, would have an evolutionary advantage. But what if mirror neurons are a product of learning? It's certainly feasible that they could be sculpted by learning, as an individual both watches and performs tasks and makes links between 'my action' and 'someone else's action'. In this scenario, motor neurons become mirror neurons – able to match performing a task with watching someone else doing it, through learning. Natural selection may also have favoured that type of learning, rather than the mirror neuron itself. There are subtle differences in the way that the mirror neuron system operates in monkeys compared with humans, and these are more easily explained if the system has arisen in each by associative learning, rather than being innate.

So what about my babies poking out their tongues? Surely that argues for an innate origin of mirror neurons? Newborn babies haven't had the experience to develop associations between what they do and what others do. Unfortunately, tongue-poking seems to be an isolated example of something where newborn babies appear to be copying adult expressions, and some scientists have argued that it's not imitation at all but a kind of non-specific response to a stimulus. It seems that the evidence for the presence of mirror neurons in newborn brains is lacking. Maybe my baby pokes his tongue out in response to all sorts of things, but I've only really noticed it when it was done in response to my own poked-out tongue.

There's also evidence that experience modifies the mirror neuron system. Pianists show more brain activation than non-pianists when watching someone else playing a piano. Ballet dancers show more activation than other dancers when watching ballet. These observations make it more likely that mirror neurons develop through learning, and depend on past experience of both watching and performing tasks.

While my baby poking his tongue out might not, after all, be anything to do with mirror neurons, he could be working on building his own mirror neuron system. He spends a lot of time looking at his own hands, moving his fingers and watching them intently, which could be a tendency that has appeared, perhaps even as a side-effect of babies being fairly helpless at birth, to help us to develop very precise control over hand movements, linking vision and muscle action. The dexterity of human hands is crucial to our ability to make tools – to create technology. Perhaps mirror neurons and the ability to link your own actions with watching those of other people are just a by-product of learning to coordinate vision and movement.

It seems that mirror neurons are more likely to be a manifestation of learning to associate tasks that we perform with those performed by others. However, these neurons are not unique to humans, so although they're fascinating, they don't explain human uniqueness. And while they are probably very important in social interaction – as well as in interpreting basic actions – they're more likely to be a product than a cause of our extremely sociable nature.

Still, mirror neurons are intriguing because they didn't fit the pattern of what neuroscientists expected to find. Perhaps we're being very naive in trying to label whole chunks of cortex as simply 'sensory' or 'motor'. Mirror neurons don't like to be pigeon-holed in that way; at the very least, they've taught us to be less narrow-minded about the brain.

Trying to pin down exactly how our brains differ from those of other primates is a tricky task. We've got huge brains, much bigger than those of even our closest primate cousins, but the problem comes in trying to work out what all that extra brain tissue is actually doing, which is why the debate over mirror neurons is so frustrating. For a moment, it looked like we might have stumbled on something really unique in human brains, something fundamental to humanness, which explained our success as a species. But mirror neurons aren't a special feature of humans, after all.

THE HUGE HUMAN BRAIN

Apes such as chimpanzees have large brains compared with most other mammals, including monkeys. But a chimp brain is just a fraction of the size of a human brain. In making these comparisons, it's really important to take into account body size. A sperm whale has a brain that weighs about 8kg, which is roughly six times the size of a human brain. But that huge brain doesn't sound so big when you take into account the whales' body mass – around 20 tonnes compared with about 70kg for a human. In other words, the brain of a sperm whale accounts for 0.04 per cent of its entire body mass, while the brain of a human weighs in at around 2 per cent.

But what does this mean? Large animals aren't just scaled-up versions of smaller ones – the laws of physics and physiology mean that proportions necessarily change at different body sizes, even without natural selection pulling parts of our bodies in different directions. We shouldn't expect brains to always represent the same percentage of body mass, regardless of the size of the animal. It has been argued that a better way of looking at brain size is to use the EQ or 'encephalisation quotient'. Across most mammals, brain mass scales to body mass to around the power of 0.75. When a mammal has a brain the size you'd expect given this scaling relationship, its EQ is 1. Chimpanzees have an EQ of about 2: their brains are twice the size you'd expect for a mammal of their body size. Humans have an EQ of more than 5 – that's five times bigger than you'd expect for our size. Over the course of human evolution, body size has certainly increased: the average body mass of an australopithecine living some three million years ago was only about 40kg, that's probably just over half your weight. You'd expect brains to have grown a bit bigger as body size increased, but there was a disproportionate growth in brain size over the course of human evolution, which is why your EQ is so large compared with other apes.

Human brains come in a surprisingly large range of sizes. Some of us have brains that are around 1 litre in volume, whereas others measure up to as much as 1.7 litres (a lot of that variation is down to body size). However, on average, the human brain is around 1.4 litres (about 2½ pints) in volume.

At this point I just have to know how big my own brain is. I've got a very accurate 3D model of it, after all, and it turns out that the volume of my brain is *exactly* 1.4 litres (according to my kitchen measuring jug). Very, very average, but at the same time, impressive compared with other mammals.

In comparison, chimpanzee brains average less than 380ml (about two-thirds of a pint), which is barely larger than the brain of a newborn human baby. My very average human brain is more than three and a half times as large, and I have around twice as many neurons in my cortical grey matter as a chimpanzee would possess in its cerebral cortex.

Human brains are not only larger than those of our closest living relatives, there has also been a reorganisation of the brain – particularly of the cerebral cortex – over the course of human evolution. This reorganisation is reflected in changing proportions of the cerebral lobes. Although the human cerebral cortex looks to some extent like a scaled-up version of a chimpanzee brain, some areas are disproportionately smaller, others larger. The occipital lobe, which is involved with vision, is relatively small. The primary visual cortex in a chimpanzee brain accounts for more than 5 per cent of the total cortex; in humans, it's a mere 2.3 per cent. The human cerebellum is also relatively small: as a proportion of total brain size, it's about 20 per cent smaller than in ape brains. On the other hand, the parietal lobes, temporal lobes and prefrontal cortex of the frontal lobe are proportionally large in humans. It's difficult to know what these differences really mean in terms of function, but it seems that natural selection has acted both on overall brain size and on the sizes of 'modules' within the brain.

We can presume (cautiously) that the areas of our brains that have become relatively larger over the course of human evolution have done so because natural selection has acted on them. Ancestors with slightly larger prefrontal areas and temporal lobes must have had a slight advantage, so that, over time, those areas grew disproportionately. What could that advantage actually be? To answer that, we need to look at the function of those expanded areas. There is, of course, a caveat, as studies of brain function show that particular tasks usually involve more than one area of the brain and there is variation between individuals. But we do know that the temporal lobe – which is proportionally 25 per cent larger in humans compared with other apes – is involved with memory and also language. This expansion of the temporal lobe seems to have happened very late in our evolutionary history – we know this from looking at fossilised skulls. The temporal lobe lies in the middle cranial fossa of the skull base, on top of the petrous temporal bone. The middle cranial fossa is about a fifth larger in *Homo sapiens*' skulls compared with archaic species, including Neanderthals and *Homo heidelbergensis* (which was probably ancestral to us and Neanderthals). Unfortunately, it's impossible to know what this means in terms of the differences in language capability between these species.

The prefrontal cortex is involved with 'higher' cognitive skills, like language, reasoning, planning and social behaviour – the functions that poor Phineas Gage lost in his encounter with the tamping iron. It's possible that the enlarged human prefrontal cortex represents the physical manifestation of a lot of what we consider to be 'modern human behaviour' – our ability to think about abstract ideas and represent them in spoken or written words, our propensity to create culture together, and our complex social networks.

The type of complex social interaction that we humans indulge in seems to depend on a particular ability called 'theory of mind', which is considered to be a keystone of human psychology. This is the ability to attribute mental states to yourself and to others, and to understand that others' beliefs and goals are different from your own. You understand that the way other people behave depends on their mental state, and you can predict how other people will behave in certain situations because you imagine what they are thinking and feeling. This ability develops in infanthood. Babies begin to take notice of others' attention, for instance, looking where a parent is looking, by about 7–9 months of age. By the time children are two to three years of age, they understand that other people's actions are goal-directed.

So, chimpanzees are extremely social animals – competitive and cooperative – and you might have thought that being able to attribute mental states to others would be very useful to such an animal. But do they have theory of mind?

Studies have produced conflicting results. One such study, published in 1996, where chimpanzees begged for food from humans whether or not those humans had buckets over their heads, suggested that chimps had no understanding that others' awareness of a situation could be different from their own. These and other studies suggest that chimpanzees only understand 'surface-level behaviour'. They are able to make predictions about how others will behave, but this is based on experience of previous behaviour and doesn't involve that mental leap to imagining what someone else is thinking and therefore being able to predict their behaviour accordingly.

Experimenters have devised tests to see if chimps understand that behaviour is goal-directed: that an individual performing an action has a particular goal in mind. They might understand that a goal is achieved by a particular action, but that doesn't mean that they understand that the actor *intended* to achieve that goal. However, if the chimpanzee is seen to understand the goal when an actor makes an unsuccessful attempt to achieve it, or has an accident in the process, that suggests something deeper is going on. If, rather than imitating an actor's unsuccessful attempt to achieve a certain outcome, the chimpanzee realises the real goal and does something to successfully achieve it, that strongly suggests that the chimp is interpreting the actor's mental state. Michael Tomasello, who studies chimpanzees as well as children, believes that the weight of the evidence has now tipped in favour of chimpanzees possessing theory of mind. The number of behavioural and contextual rules that they would have to be applying, in order to produce similar results in the many experiments that have attempted to test their appreciation of goal-directed behaviour, means that surface-level understanding is unlikely.

Another indicator that chimpanzees possess theory of mind is their awareness of others' attention. They will follow gaze (more often following head direction rather than gaze direction, but the latter may be a particularly human thing to do, helped by the anatomy of our eyes where the white sclera makes following gaze very easy). They look for the object of another person's attention, just as human infants do. In studies where chimpanzees are competing for food, they will behave in a way which is hard to explain if they don't have theory of mind. For instance, when other chimpanzees are within sight or hearing, they will try to conceal their approach to a hidden source of food.

The wealth of sophisticated studies that have been carried out in the last decade or so point to one reasonable conclusion: that chimpanzees have a theory of mind similar to our own – as Michael Tomasello put it, 'they understand that others see, hear and know things'. Nevertheless, the parts of our brains which are thought to be involved in social interactions are disproportionately large.

We can trace the expansion of the human brain, and to some extent its different modules, by looking at fossil skulls of ancient hominins. Our very early ancestors' brains were more or less chimpanzee-sized. The 6–7-million-year-old hominin called *Sahelanthropus tchadensis* ('Toumai' to his friends) had a brain of around just 350ml in volume. The average brain size of *Australopithecus afarensis* (a hominin species which lived in Africa three to four million years ago and which includes the famous specimen of 'Lucy') was around 440ml, giving an EQ of about 2.5 – not much bigger than other living apes today. Although the brains of *Homo erectus*, a species which lived in Africa and Asia from around two million years ago, seem to have ranged widely in size, they averaged around 910ml, with a mean EQ of 3.7. But the really impressive increase in brain size came quite late in human evolution. From around a million years ago, with species like *Homo heidelbergensis*,

Neanderthals, and us – *Homo sapiens* – arriving on the scene, we see average brain sizes of much more than a litre, and EQs rising to modern levels of 4–5.

Of course, growing large brains had a significant knock-on effect for brainboxes: human skulls have had to expand to accommodate a bigger brain. The braincase dominates the human skull in a way that it just doesn't in other species, even chimpanzees, which still have relatively large brains compared with other species. But the skull has also had to change shape to hold a bigger temporal lobe. One of the implications of this might have been to cause an angulation of the skull base. This angle is a special characteristic of human skulls – helping the skull to accommodate a very large brain.

Here's a two-dimensional example which I think helps to explain this idea: imagine a Spanish fan, opened to its full extent, with the outer guardsticks hinged open to 180 degrees, to lie in straight line. You could make the fan larger by adding more leaves to the fan, and opening it perhaps another 40 degrees. It wouldn't sit as neatly in your hand, but what this oversized fan illustrates is how flexing the base of the skull would allow more brain to be fitted inside the skull above it. The flexing of the base of the skull has another knock-on effect, in that it pulls the facial part of the skull in, tucking it under the front of the braincase.

While growing large brains has exerted a powerful influence on the shape of our skulls, this is a very recent footnote to their evolution, an important bit of anatomy which originated with the earliest vertebrates. That ancient story is one that was woven into the embryological development of your own skull.

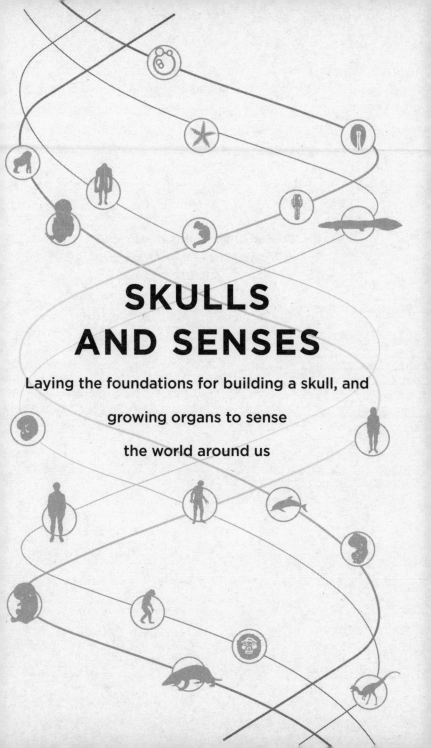

SKULLS
AND SENSES

Laying the foundations for building a skull, and

growing organs to sense

the world around us

'Ut imago est animi voltus sic indices oculi'
('The eyes are the window to the soul')
CICERO

NEURAL CREST AND THE ORIGIN OF THE SKULL

I t's week three of your embryonic development and your neural tube – which will eventually become your brain and spinal cord – is forming. But as the neural folds rise up and then push together to form the neural tube, cells at the apex of each fold become restless – they are about to make a break for it. As the folds fuse, the crest cells weigh anchor and set sail, migrating through the embryo to reach their many destinations. They're such an important population of cells that they're sometimes called the 'fourth germ layer': there are ectoderm, mesoderm and endoderm, and also neural crest. These cells spread through the body and form a diverse panoply of tissues, including parts of the adrenal glands, membranes around the brain and bits of your skull – specifically, the bones which form your face. All vertebrates – that's any fish, amphibian, reptile, bird or mammal you care to mention – have these neural crest cells in their embryos. Skulls are just as fundamental to being a vertebrate as having a spine, and neural crest cells are essential for making a skull.

So how did neural crests – and skulls – suddenly spring into being when the first vertebrates evolved? This is evolution we're talking about, not divine creation, and as a couple of developmental biologists have observed: 'structures . . . do not simply arise from the dust of the earth'. This is where genetics comes to the rescue. The little lancelet is a chordate, but not a vertebrate. It lacks both neural crest tissue and a skull, and there's no reason to believe that either of these existed in ancient pre-vertebrates like *Haikouella* either. We can't look at the DNA of *Haikouella* – it's long since disappeared – but we *can* look at the genes of its living counterpart, the lancelet. Although there are 530 million years between living lancelets and *Haikouella*, the lancelet really does look like a 'living fossil', and it's unlikely that any ancestors of lancelets acquired and then lost vertebrate characteristics like neural crest tissue, and a skull. It's much more likely that the neural crest, along with skulls, never featured in all those generations between Cambrian pre-vertebrates and lancelets. This means that the genes for neural crest should be completely absent from the lancelet genome. If we want to find out which genes make a vertebrate a vertebrate, the real value of the lancelet lies in precisely what it lacks.

To find out what specific genetic changes are responsible for 'switching on' neural crest cells, or for making a skull, it's best to compare the lancelet with a relatively primitive vertebrate. (We don't want to get confused by genetic changes which led, later in vertebrate evolution, to novelties like limbs, for instance.) The first vertebrates were jawless fish or 'agnathans' (literally 'jawless' in Greek) and there are two groups of these alive today: hagfish and lampreys.

Lampreys have achieved a strange sort of fame by having constituted the final, fatal meal of Henry I, according to the chronicler Henry of Huntingdon. Henry I was the son of William the Conqueror, and became king in 1100 after his older brother was killed in a hunting accident. He reigned for 35 years, and at the age of 68, apparently in direct contradiction of his doctor's orders, he ate a 'surfeit of lampreys', took ill the next day, and died within the week. It seems like such an odd thing to eat, but lampreys were quite a delicacy in ancient Rome and continued to be enjoyed by Medieval aristocrats. On Twitter, I politely enquired if anyone had consumed lamprey and knew what it tasted like. Respondents wrote back variously: 'a lot like Moray eel', 'muddy', 'chicken', and 'like soap', so I'm none the wiser and unlikely ever to find out as river lampreys are now an endangered species in the UK. Although not in Japan, it seems. My friend, the historian Neil Oliver, replied with a splendid tweet: 'Believe it or not, they were part of a banquet I shared with Lord Satsuma, a Shogun, at his home in Kagoshima.'

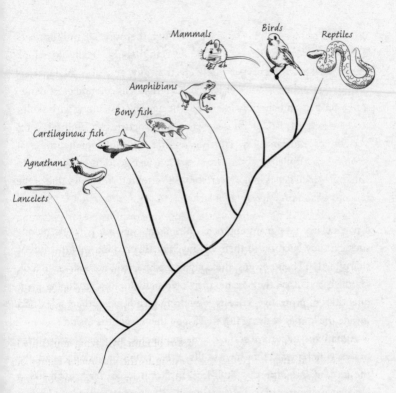

Mammals

Birds

Reptiles

Amphibians

Bony fish

Cartilaginous fish

Agnathans

Lancelets

A family tree (with earliest ancestor at the bottom) of vertebrates, their close cousins, agnathans (including lampreys) and their more distant cousins, lancelets

Some lampreys live in rivers, and others in the sea, and they look like eels: they are long (up to a metre long) and flexible, and have no scales. But unlike eels, lampreys have no jaws. A lamprey's mouth is like a miniature version of the sarlacc's mouth at the bottom of the pit in *Return of the Jedi*. (In fact, it's so similar that I suspect the set designer must have seen a lamprey and thought it nightmarish enough to provide the model for this loathsome alien.)

Being jawless, agnathan fish are somewhat limited in what they can eat. Early jawless fish were filter feeders, but modern lampreys have developed

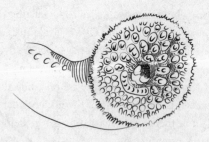

A lamprey

a particularly repugnant approach to dining. They are parasitic blood-suckers: they latch on to their victims with their round mouths, full of sharp teeth. (Their cousins, the aptly named hagfish, are also quite foul, ghoulish creatures: they escape their own predators by producing great quantities of protective, viscous slime from glands along their sides, and invade the bodies of dying fish to feast on them from the inside.)

Agnathans may have some disgusting habits, but they're still verte-brates, which means they have skulls. At least, they have skulls as adults; the larva of a lamprey is skull-less, in fact it looks very much like a lancelet. So lamprey larvae have (just like the lancelet) gill slits, a hollow nerve cord running along the back, a notochord, segmented muscles and a tail which extends beyond the anal opening – everything you'd expect in a chordate. But when lamprey larvae metamorphose into adults, they grow an internal cartilage skeleton including vertebrae and a braincase: the hallmarks of vertebrates.

Comparing the lancelet and lamprey genomes, differences shine out. It's like comparing two 'spot the difference' pictures. The differences in the lamprey's genome enable it to undergo that incredible transforma-tion, from something that looks like an unassuming pre-vertebrate to a creature with a spine and a skull.

Just as parts of anatomy don't suddenly spring into existence, out of nowhere, neither do new genes. They have to come from *somewhere*, and new genes usually appear as duplicates of existing ones, and they occur due to mistakes in copying DNA. Scrutinising genomes, it becomes clear that large chunks of DNA have become duplicated in the vertebrate lineage. For instance, mammals have four clusters of the pattern-generating genes known as *Hox* genes, while the lancelet has just one. The lancelet is a living relic of the ancient ancestors of vertebrates, who relied on their single set of *Hox* genes, before those large-scale gene duplications occurred.

When a duplicate gene appears in a genome, several things can happen. Sometimes one version is essentially surplus to requirements and may degenerate or even disappear entirely. But there's another, more interesting, possibility: one gene may continue to fulfil its old function while the duplicate is free to do something new. Genes may also have more than one function to start with; for instance, they might get switched on at different times during development, performing a different job each time. So a third possible fate for duplicate genes occurs when you get two versions of the same gene and these different jobs end up getting shared out between them. Both genes become necessary within the genome, and neither will degenerate, but they can evolve off in different directions, eventually taking on new functions.

When a developing lancelet is making its neural tube, there's a group of cells which look *almost* like the neural crest cells in vertebrate embryos – they turn up in a similar position and have some similar genes switched on. These not-quite-neural-crest cells even move away from the neural tube – but they don't move very far. The difference between these cells and the true neural crest cells of vertebrates seems to come down to a particular set of duplicated genes. Among them is one called FOXD3. Lancelets only have one copy of the FOXD gene, whereas vertebrates have four or five. The lancelet FOXD gene is silent near the developing neural tube, but in lampreys and other vertebrates one of the copies, FOXD3, is working hard at the top of the neural fold. This seems to be one of the genes which 'tells' the neural crest what it is and kick-starts its migratory march through the embryo.

Millions of years ago, a few enthusiastic, duplicated genes, taking on a range of new roles in the developing embryo, led to the evolution of vertebrates, complete with neural crest and skulls. And, much more recently, those same genes were busy in the developing embryo which became you. Cells at the crest of your neural folds responded to that genetic signal, which goes all the way back to your earliest vertebrate ancestors. Each of those cells grew small protrusions, very similar to the pseudopodia (Greek for 'false feet') of an amoeba, and then crawled off through the thronging mass of cells in the middle layer of your embryonic germ disc.

Some of these cells only crawled a short distance, staying close to the spinal cord and forming knots – ganglia – of sensory nerve cells. Others crawled off into the neck and lay in wait to become the hormone-secreting cells of your thyroid gland. Some made their way to where the mouth would form, intent on helping to form your teeth. Others ended up in a septum between the great arteries leaving the heart. Some made it into the core of your adrenal glands, where they would eventually be responsible for producing the hormone adrenaline. Still others filled themselves with fat and wrapped themselves around the extensions from nerve cells, forming an insulating sheath around nerve fibres. Some of the cells moved far away from their birthplace, ending up in your skin, and making pigment there. Many of them crept into your developing head, where they began to lay down the foundations of your skull. Without neural crest cells, you wouldn't have a face.

Pick up any self-respecting embryology textbook and you'll find a long list of all the tissues in the adult body which start off as neural crest. But uncovering the *curricula vitae* of all those neural crest cells was fraught with difficulty – both academic and technical.

In 1893, Julia Platt, a woman who spent nine years studying embryology for a PhD at both Harvard University and Freiburg, in Germany, published a paper making the controversial claim that in a salamander embryo some of the cartilages in the head were formed from neural crest cells. At the time it was an outrageous claim, because all cartilage and bone was thought to come from the middle, mesoderm layer of the embryo. Everyone knew that ectoderm (which is where neural crest comes from) made epidermis and neural structures. To suggest anything else was faintly ridiculous. What was this Californian woman on about? A storm of destructive criticism built up in the embryological literature. One embryologist even suggested that Miss Platt must have prepared her specimens badly. Eventually, Julia Platt would be proved right, but too late to save her own career in science, which ended with the completion of her PhD in 1898.

Part of the reason why Julia Platt's claims were doubted by her contemporaries was that she was focusing on quite subtle differences in appearance between cells in the developing embryo, so subtle that some doubted these differences existed at all. Julia Platt studied thin sections of salamander embryo specimens under the microscope, and she believed that she could tell the difference between cells that had originated from ectoderm, compared with mesoderm and endoderm. This does sound fraught with difficulty, but it was a method that other embryologists had also used: the ectodermal cells (including neural crest cells) were smaller than mesoderm cells and contained granules of brown pigment. But these studies relied on looking at specimens of different embryos across a range of ages in order to reconstruct the migration of the neural crest cells. How could the embryologists be *sure* that they were seeing the same cells, from one embryo to another?

Tracking cells got easier with the development of a technique called 'fate-mapping'. This sounds somewhat prophetic but there are no crystal balls involved. Instead, embryologists inject dye into the living cells of an embryo and then watch to see where they end up. In the 1920s, embryologists were staining migrating cells blue or red. By the late 1970s they had a range of fluorescent dyes, and by the 1980s embryologists were creating genetically altered embryos with cells which would effectively 'dye themselves' various fluorescent colours (pioneering the techniques which would eventually make the 'Brainbow' possible). Now it was possible to tag the neural crest cells and watch them migrating.

Embryologists observed neural crest cells detaching themselves and streaming forwards in the head, ready to start laying down the foundations of the front of the skull, including the frontal bones and the upper and lower jaws. Now all they had to do was turn into bone.

MAKING A SKULL

When you were still just a three-layered embryo, cells in your middle, mesoderm layer, together with migrated neural crest cells at the head end, formed a special type of embryonic tissue called mesenchyme. Mesenchymal cells still haven't made their minds up about exactly what they want to be when they grow up. They are still undecided – undifferentiated. Most of them will commit to a particular destiny quite soon, as bone, cartilage, muscle, or blood cells, for instance. But a few undifferentiated stem cells stick around much longer, they're still there in

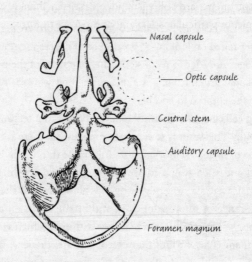

The chondrocranium (cartilage base of the skull) of a human embryo

your adult body, in marrow and in fat tissue, as a source of replacement cells. The whole field of regenerative medicine is based on the potential of using these 'mesenchymal stem cells' to repair damage.

For many cells that are eventually going to form bone, the first step is to turn from mesenchyme into cartilage. In what will become the base of the skull, mesenchyme forms itself into islands of cartilage, before joining up and then transforming into bone. This way of making bone, by first creating a cartilage 'model', is the most widespread form of ossification in the embryo: the spine, ribs, sternum and all the limb bones apart from the clavicles also ossify in this way.

In the developing skull, we see yet another example of what Darwin called the 'law of embryonic resemblance' (1859). As adults, our skulls look quite different from those of most other mammals. But when you compare the cartilage model of the base of the embryonic human skull with the developing skull of other mammals, the similarities are striking. The skull base is made up of three pairs of capsules which contain the nasal cavity, the eye and the ear. In many other mammals, these compartments are still easily identifiable in the adult skull: the nasal capsule, in particular, sticks out at the front to form a snout. The

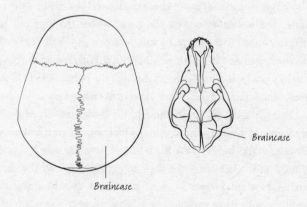

The top of a human skull and a dog skull

braincase is small and neat and sits over the auditory capsules, which contain the workings of the ear. In contrast, our braincases have grown so large that if you look at the top of an adult human skull, you see nothing else. Our eyes are tucked under the front of the braincase; our external noses are very short and not at all snout-like, and the whole of the nasal cavity is secreted away under the braincase too.

Most of the bones in your skeletons developed in the same way as the base of the skull, with mesenchyme first turning into little cartilage models, which later transform into hard bone. But the vault of the skull, forming the top of the braincase, together with parts of the facial skeleton and the clavicle, are all unusual in ossifying straight out of mesenchyme.

In the mesenchyme lying over the developing brain, bone cells differentiate and start to produce spicules of bony matrix, containing a mixture of protein and bone mineral. The spicules gradually radiate out from a centre of ossification but the edges of the bones are still slightly separated at birth, with a fibrous membrane forming a connection. Where two bones come together, the narrow seam is known as a suture. Where several bones meet each other, the membrane-covered opening is called a fontanelle. These are the 'soft spots' of the baby's head – a larger one at the front, where the frontal bone meets the parietal bone, and a smaller one at the back, where the parietals meet the occipital bone. These fibrous joints between the skull bones allow them to overlap just a little during childbirth, slightly easing the passage of the baby's head (9cm in diameter) through the pelvic outlet (10cm in diameter) and resulting in some newborns with very odd-shaped heads. Over the next day or so, though, the bones usually settle back into position and the baby's head starts to look more round and normal. The fibrous sutures between the flat bones of the vault stay open as the child's head grows, allowing the skull to expand. In fact, the sutures are crucial to skull growth – they are where growth *happens*, as the membranes in these gaps add new bone tissue to the edges of the skull bones.

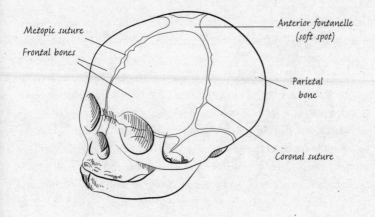

Metopic suture

Frontal bones

Anterior fontanelle (soft spot)

Parietal bone

Coronal suture

The sutures and fontanelles in a newborn baby's skull

SKULL SHAPES

Babies' skulls are very malleable. They can be accidentally or deliberately deformed relatively easily. Deliberate deformation of young skulls was practised in ancient Egypt, Greece, Peru and Australia. Weird head shapes seem to have been used as markers of social identity, showing that an individual belonged to a particular group, or as indicators of status within a group – marking out someone as a member of a social elite or warrior class. The Egyptian pharaoh Akhenaten's head appears to have been deformed by a pathology, but sculptures of his queen, Nefertiti, suggest that she also had an elongated, conical head. The mummies of Tutankhamun and his mother also show evidence of some conical deformation. Depictions of the royal family show them all with extremely long heads, and it looks like this was more than just artistic licence – they really did have deliberately deformed, cone-shaped heads.

Early Mesoamericans bound their babies' heads between two pieces of wood – one at the forehead, the other at the back – flattening the head

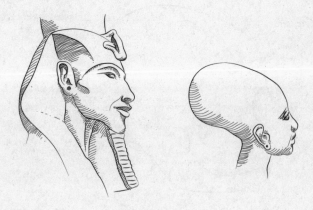

Statue of Akhenaten (left) and sculpted head of one of Akhenaten and Nefertiti's daughters (right)

in front and behind, and making it grow taller. Historical cultures in Europe, Asia and Africa have also encouraged the binding of babies' heads to produce a desirable, conical skull. A recent review paper in the academic journal, *Clinical Anatomy*, reminded clinicians that odd-shaped baby skulls might still be produced by family members employing various techniques to mould them. Remarkably, as recently as 1994, a book was published with the title *Baby Beautiful: a Handbook of Baby Head Shaping*.

Incidental deformation of babies' skulls was common among indigenous American communities, who strapped their infants to baby-carriers called cradle boards. This led to a 'posturally produced' flattening of the back of the skull – the occiput. Since the early 1990s, a similar effect has been noticed following advice to parents, in the US and the UK, to put their babies to sleep on their backs. The 'Back to Sleep' campaign has been a great success, reducing the rate of babies dying from sudden infant death syndrome (SIDS), but while the campaign also resulted in a reduced frequency of babies with flattened foreheads from sleeping on their fronts, we now have more babies with flattened occiputs. Fortunately, the occipital flattening which this posture produces seems to be a temporary effect.

None of these skull deformities, caused by deliberate or incidental pressure to different areas, seem to cause any issues for the developing brain inside the skull. However, real problems arise if some of the skull sutures close too early, restricting the growth of the skull. Early closure of the sagittal suture means that the skull cannot grow any wider, and so it pushes out at the front and back, becoming long and narrow. Premature fusion of the coronal suture prevents any more growth between the frontal and parietal bones, making the skull very short and tall.

Although bone seems such a rigid tissue – and mature bone is – pressure inside the skull will influence the shape of the vault bones as they grow. This is seen when the sutures fuse too early, but also when the intracranial pressure is abnormally high. In hydrocephalus (from the Greek for 'water' and 'head'), an excess of cerebrospinal fluid (CSF) inside the skull can create pressure which causes the skull to balloon. At its roots, this pathology goes back to the formation of the brain from the hollow neural tube in the embryo. As the brain develops, the space

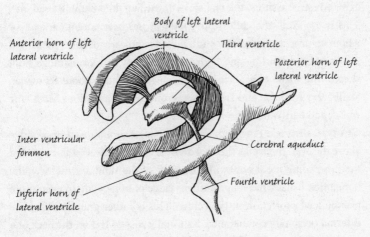

The ventricles inside a normal brain

inside becomes elaborated into a series of interconnecting chambers or ventricles, filled with CSF.

Some of the channels between the ventricles, or out into the spinal cord, are very narrow, and a blockage in these leads to a build-up of CSF. It is possible to treat this condition surgically: a tube can be inserted into the brain to drain away excess fluid to another body cavity, usually in the abdomen, or a small opening can be made in the base of the brain, allowing the CSF to escape.

In 1999, a paranormal researcher called Lloyd Pye claimed that a strange skull, apparently found in a mine in Mexico in 1930, was probably the offspring of a human woman and an extraterrestrial. In a YouTube video, Pye describes the 'Starchild Skull' as 'having virtually nothing in common with a human skull.' Pye quotes one craniofacial surgeon in Canada as saying that he had not seen any comparably deformed skulls in his career, but it seems extraordinary to assume that this means an extraterrestrial explanation is *more* likely. The skull is missing the maxilla or upper jaw, which perhaps makes it look even more peculiar to untrained eyes. It's certainly unusual, but it looks very much like the cranium of a child who died from hydrocephalus. Rather than an exciting tale about extraterrestrial visitors, the real story of the Starchild Skull is a sad one. This is the skull of a child whose identity has been forgotten, and for whom an untreated medical condition led to an early death.

While head-binding and hydrocephalus may produce strange skull shapes, the shape of a normal skull can reveal much about its owner. Skulls vary in shape and size between different populations across the world, and between sexes.

A typically male European skull will have a pronounced brow ridge above the eyes. In profile, there's a definite indent between this ridge and the protruding nasal bones. At the back of the skull, on the occipital bone, men usually have a prominent ridge of bone with an even more pronounced projection in the centre. This is rather grandly called the external occipital protuberance. It's usually easy to feel on the back of a man's head. It may even be visible, if he's bald or has a shaven head. The

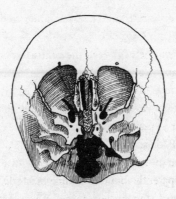

The 'starchild' skull

ridge has a curve on each side, declining towards the midline protuberance, forming a shape like a cartoon seagull. Below the ear, an inverse pyramid of bone called the mastoid process projects down. This is easy to feel, just under your earlobe, and is much larger in a man than in a woman. The mastoid processes and the occipital protuberance are both areas where muscles attach to the skull. The mastoid process forms the anchor for a long strap of muscle which runs diagonally down the neck – the sternocleidomastoid muscle. When you turn your head to the side, this muscle stands out on the opposite side of your neck – in fact, it's the contraction of this muscle which is turning your head. The occipital protuberance, together with the seagull-wing ridges either side of it, give attachment to muscles at the back of the neck, including a large, kite-shaped muscle called trapezius.

A woman's skull is usually less robust, or, to put it another way, more gracile. All of these features will be smaller or more subtle. The brow ridge may be practically non-existent: a gently curving forehead, formed by the frontal bone, descends and curves out into the bony part of the nose. The mastoid process is small and neat, and the back of the occiput is often completely smooth and round, with barely a ripple where the muscles of the neck rise up to attach to it.

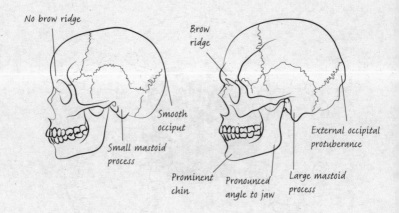

Sex differences in human skulls: female (left) and male (right)

When I'm looking at archaeological skeletons and trying to deter-mine the sex of an adult, a well-preserved skull is a gift. Even if the pelvis is missing or in fragments, the skull will help me decide whether that person was male or female. The teeth will give me further clues, helping me to determine the age of the person when they died. An immature set of teeth is easy to age: teeth form and erupt at different times, in a fairly strict sequence. Twenty milk teeth fill the mouth by the time a child is three years old; in each quadrant, or half-jaw, there are two incisors, one canine and two molars. Three years later, the child will start to shed those baby teeth, and permanent teeth will take their place: 32 of them at the final count, including the third molars or wisdom teeth (unless you are among the 10–20 per cent of the population who fail to grow a third molar). The baby incisors and canines are replaced by larger incisors and canines and the baby molars are replaced by two-cusped adult premolars, then the three adult molars fill the new space in the enlarging jaw, behind the milk teeth. Very roughly, the first perma-nent molar erupts at six years, the second at twelve years and the third at eighteen years.

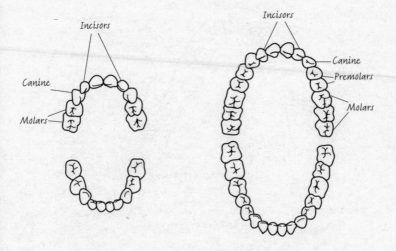

Deciduous or milk teeth (left) and permanent or adult teeth (right)

Once all the teeth have erupted into the mouth, it becomes more difficult to age a skull. You can get a very rough idea of whether you're looking at a young or an elderly adult from the sutures between the flat plates of the skull, though. These joints, which are wide open and separated by membranes at birth, become very narrow and interlocking during childhood, but the bones are still separated by a small amount of fibrous tissue. As you get older still, the sutures close and the bones fuse together. Unfortunately, unlike tooth eruption, there's no set programme for suture closure, so it remains only a rough guide to age. But the teeth can be more useful, especially in archaeological human remains.

We eat such a refined diet today that our teeth wear down more slowly than those of our ancient ancestors, whose bread was full of fine grit from quern stones. Although tooth enamel is the hardest substance in the human body, designed to withstand heavy grinding, it will eventually wear away, exposing the dentine core of the tooth, so the degree of tooth wear can be a useful indicator of age. Dentine will be exposed

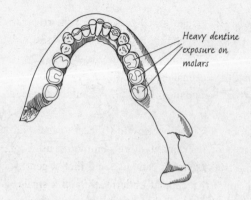

Heavy dentine exposure on molars

Mandible from a Medieval burial ground showing heavy dental attrition (suggesting an age of 35 to 45 years at death)

first on the tips of the teeth, along the edges of the incisors and on the cusps of premolars and molars. As the years pass, teeth are ground down further, until cusps are obliterated. A heavily worn molar has a flattened grinding surface of dentine, with just a thin rim of enamel around the edge. The tooth may be ground down so far that all the enamel is lost, and eventually worn down as far as the roots. As a modern-day human, my diet is much kinder to my teeth than it would have been had I lived more than a few centuries ago. Nevertheless, when I brush my teeth and look at my open mouth in the mirror, I can see that, while my molar cusps are bearing up well, there's a thin line of dentine exposed on the tips of each of my incisors. Slowly but surely, my teeth are wearing down too. You might want to check yours.

HEARING

The skull and teeth are incredibly useful to a physical anthropologist who is trying to determine the age and sex of a person from just skeletal clues. Whenever a particularly well-preserved skull arrived in the bone lab, I knew I had a good chance of finding out the sex and age of its owner, but I also used to find it almost impossible to resist the temptation of looking for the smallest bones in the human body. These tiny bones wouldn't really help me in my analysis, but they were fascinating, just the same.

A very cautious exploration through the ear canal with a dental curette (the type of instrument more commonly used for scraping calculus off the teeth of living patients), and then a gentle tapping of the skull with one hand cupped underneath, and a small pile of dry dirt would fall out into my palm. Sorting through the grains of dirt with the tip of a finger, I would look for the tiny ossicles: three of them: the malleus (hammer), incus (anvil) and stapes (stirrup). The malleus measures about 9mm in length, while the minute stapes is just 3mm long. It seems so extraordinary that such diminutive structures would ever be preserved, and yet they often are, and perfectly so. In life, these bones are connected to each other by minute joints and

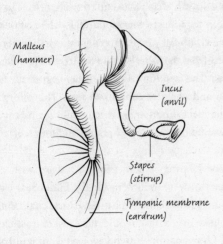

Malleus
(hammer)

Incus
(anvil)

Stapes
(stirrup)

Tympanic membrane
(eardrum)

Auditory ossicles: the smallest bones

ligaments. This system is designed to convert waves in air to waves in fluid.

The eardrum or tympanic membrane lies at the inner end of the external auditory meatus, completely sealing it. If you use an otoscope to look into a healthy ear canal, in a living person, you can see the tympanic membrane at the end – a glistening, slightly pearly and almost translucent membrane. You can even see the malleus attached to the back of the membrane. Sound waves entering the ear cause this delicate membrane to quiver, and then the chain of bones carries those vibrations across the space of the middle ear, to the cochlea.

The cochlea is a snail-shaped, fluid-filled cavity set deep inside the temporal bone, in the base of the skull. Close to the cochlea, there are other fluid-filled tubes inside this bone: three C-shaped, semicircular canals. They sit at 90 degrees to each other – perfectly positioned for detecting acceleration and deceleration in three planes. In both the cochlea and the semicircular canals there are cells with tiny 'hairs' which detect movement in the fluid inside these tubes.

Vibrations in the fluid inside the cochlea kick-start an electrical impulse in the hair cells, which is picked up by nerve cells. The fibres of these nerve cells join together to form a bundle – the cochlear nerve. This unites with the vestibular nerve, carrying information about position and acceleration from the other parts of the labyrinth, including the semicircular canals. The resulting vestibulocochlear nerve exits the skull on the inside and runs into the brainstem. The information is relayed from one neuron to another, up through the brainstem and into the brain itself, eventually ending up at the auditory cortex in the temporal lobe.

This basic kit for hearing is something we share with all other mammals. Whether you were to look inside the middle ear cavity of a mouse or an elephant, you'd see the same three-ossicle chain, linking the eardrum to the cochlea. In contrast, reptiles have just one ossicle in their ear, and the story of how mammals (who evolved from reptiles) ended up with three ossicles is an extraordinary one, which we'll return to later.

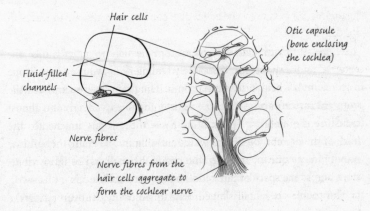

Cross section through the cochlea, cutting through the three fluid-filled channels which spiral around the central axis or modiolus. The image to the left shows an enlarged section, with hair cells

The ear is such an intricate, complex and apparently delicate system – from the translucent thinness of the eardrum to those tiny ossicles, and then those microscopic hair cells inside the cochlea – it seems extraordinary that problems with hearing aren't more common than they are. Hearing loss ranges from total to partial, and may be genetic and present from birth, or acquired. All of us suffer hearing loss to some degree as we age, losing the ability to hear high-frequency sounds. Apart from that inevitable decline, there are many other causes of acquired hearing loss, including infections like measles, mumps and meningitis, some medications, as well as some solvents and pesticides. But you can also lose hearing through exposure to very loud sounds, and this accounts for around half of all acquired hearing loss. The hair cells in the cochlea are extremely fragile, and a loud noise creates vibrations in the fluid which lays them flat, like a field of ripe wheat levelled by a strong wind.

SMELLING

There are so many elements of anatomy which we utterly rely on but which seem shockingly delicate and fragile when you uncover them in the bone lab or the dissection room. The brain itself seems so well protected inside its bony case, but yet it is possible to damage it without breaching the integrity of the skull. Even without any impact to the head, a fast acceleration or deceleration can injure the brain, the sudden movement causing the brain to shift inside the skull and collide with it, becoming bruised. Although brain contusion usually heals on its own, the symptoms are unpleasant, including headaches (unsurprisingly), dizziness and nausea. Another, long-term, result of traumatic brain injury may be a loss of the sense of smell: anosmia. If you look inside a skull, it's easy to understand why the olfactory nerves, which convey information about smell from the nasal cavity to the brain, are particularly at risk in traumatic brain injury. At the front of the skull, the frontal lobes of the brain rest on a shelf of bone, which also forms the roof of the eye-sockets. Right in the middle, between the eye-sockets, there's a narrow piece of bone which is peppered with tiny holes. This is the cribriform plate of the ethmoid bone. ('Cribriform' means 'sieve-like' in Latin; 'ethmoid' means 'sieve-like' in Greek. So, the sieve-like plate of the sieve-like bone. It seems like overkill, but the ethmoid does have other, quite un-sieve-like parts as well, although it obviously gets its name from this perforated plate.) The bottom of the cribriform plate forms the long, narrow roof of the nasal cavity. The lining of the nasal cavity here contains specialised nerve cells which have tiny projections bearing proteins (olfactory receptors) in their membranes. These proteins recognise particular odour molecules; when a molecule binds to a receptor, the receptor changes shape and kick-starts an electrical impulse. That impulse then travels up through the nerve fibres, which are bundled together as the olfactory nerves, through the cribriform plate. Just above the cribriform plate, the olfactory nerves enter a structure called the olfactory bulb, where they synapse with a second set of

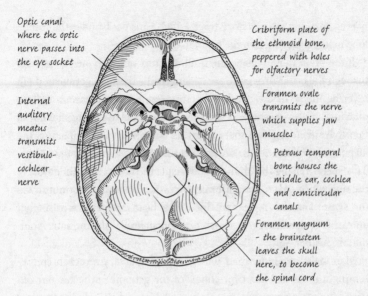

Optic canal where the optic nerve passes into the eye socket

Cribriform plate of the ethmoid bone, peppered with holes for olfactory nerves

Internal auditory meatus transmits vestibulo-cochlear nerve

Foramen ovale transmits the nerve which supplies jaw muscles

Petrous temporal bone houses the middle ear, cochlea and semicircular canals

Foramen magnum – the brainstem leaves the skull here, to become the spinal cord

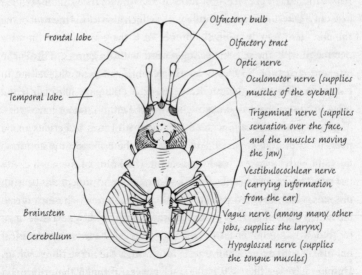

Frontal lobe

Olfactory bulb

Olfactory tract

Optic nerve

Oculomotor nerve (supplies muscles of the eyeball)

Temporal lobe

Trigeminal nerve (supplies sensation over the face, and the muscles moving the jaw)

Vestibulocochlear nerve (carrying information from the ear)

Brainstem

Cerebellum

Vagus nerve (among many other jobs, supplies the larynx)

Hypoglossal nerve (supplies the tongue muscles)

Top: A skull with the top removed to show an inner view of the skull base, showing the holes or foramina through which nerves and vessels pass. Bottom: The undersurface of the brain showing the cranial nerves emerging

nerves, forming the olfactory tract, which runs back under the frontal lobe before plunging into the brain itself.

If the brain moves inside the skull, the tiny olfactory nerves passing through the cribriform plate are endangered – they can get sheared off, severing the link between the olfactory receptors in the nose and the olfactory bulbs. A study of patients with traumatic brain injury in Pennsylvania found that more than half had some loss of olfaction, but 40 per cent of those were unaware that they had lost their sense of smell. This seems remarkable, but of all our special senses, smell is probably the one which causes the least problem when it's lost. That's not to say our sense of smell is redundant, but, certainly in comparison with other animals, it's something in which we've disinvested during our evolutionary history.

Mammals have more than 1000 genes for olfactory receptors (representing a sizeable 3 per cent chunk of the genome), whereas our old friend the sea lamprey has only around 60. Not surprisingly, the lamprey can only detect a limited number of different smells. However, even though there is only a small number of these genes in the lamprey compared with any self-respecting mammal, those genes are similar in size and structure. So this is another example of gene duplication in evolution. This time, the duplicate genes are taking on subtly different roles from the original, encoding slightly different olfactory receptors. Through gene duplication, the family of smell genes has undergone a massive, explosive expansion in most vertebrates since the common ancestor with lampreys, reaching its apogee in mammals.

I've often enjoyed playing 'hide and seek' with my dog, Bob, using an old tennis ball, but he doesn't *seek* out the ball so much as smell it out. Some 500 million years of evolution (that's how far back you have to go to find a common ancestor with a lamprey) have equipped most mammals with remarkable noses. Most mammals, that is, except us. Humans have less than 400 active smell genes. Around 400 more are still present in our genomes, but have degenerated so much that they have become inactive, mutating to the point they can no longer be 'read' and translated into proteins. Dogs – and indeed mice – have three times as many active smell genes as humans, which is perhaps why Bob is so good at sniffing out his favourite tennis ball, no matter where I hide it. I can smell this tennis ball if I put it right up to my nose, certainly (and it's not exactly a pleasant aroma), but I haven't a snowball's chance in hell of smelling it from over on the other side of the room.

So if humans have so many fewer active smell genes, and so many more inactive smell genes than most other mammals, when did this deliberate neglect of a finely honed sense begin? It turns out that it's not just humans who have such paltry numbers of smell genes. Comparing samples of genomes, one group of geneticists thought that they had found the moment when the rot had really set in. It looked as though apes, including humans, and together with many Old World monkeys, such as langurs, baboons and colobus monkeys, had significantly larger proportions of inactive smell genes compared with New World monkeys and lemurs. These two groups of primates are also different in the way they look at the world. Lemurs and lorises, together with most New World monkeys, have just two types of colour receptor in their eyes; Old World monkeys and apes have three, giving them so-called trichromatic colour vision. In other words, primates which have developed what we would think of as 'colour vision' have disinvested in their sense of smell.

It's a neat story, but unfortunately, more detailed analyses of entire genomes have failed to support it. Instead, the loss of smell genes seems to be a very general tendency among primates. Most primates appear to have between 300 and 400 active smell genes. And in fact, although we tend to think of ourselves as particularly lacking in the olfactory department, humans turn out to be prodigies among primates, with more active genes than chimpanzees, orangutans, marmosets, macaques and bush babies.

Although the evidence no longer seems to support a disinvestment in the sense of smell with the evolution of *colour* vision, it is still true that primates as a group have fewer smell genes than other mammals. Along with this reduction in the genetic basis of smell, most primates (apart from strepsirhines, like lemurs) have much shorter snouts than most mammals, and much smaller olfactory bulbs. While smell and hearing are dominant senses for many mammals, primates as a group seem to have specialised in sight and touch instead. So the story of smell has collided with the story of vision. But to understand the origins of our own eyes and vision, we need to look at more distant relatives and go much further back in time.

SEEING

Eyes like ours have been around for at least 500 million years – vertebrates had eyes right from the start. The first vertebrates, similar to living lampreys, may have been jawless, but they had eyes. In fact, the lamprey's eyes are very complex – similar to any other vertebrate eye. They are camera-like, with an iris and an internal lens, and have six muscles to move the eyeball around – just like our own eyes. It's almost as though the vertebrate eye simply sprang into being, fully formed. That's a bit of a challenge for evolutionary theory, which predicts that structures should change gradually over time, so it is no wonder that eyes have been jumped on by creationists. Even Darwin admitted eyes were difficult to explain, writing in *On the Origin of Species:*

> To suppose that the eye, with all its inimitable contrivances for adjusting the focus to different distances, for admitting different amounts of light, and for the correction of spherical and chromatic aberration, could have been formed by natural selection, seems, I freely confess, absurd in the highest possible degree.

Creationists love to quote this sentence of Darwin's – and stop there. But this sentence is just a rhetorical device, luring us in. Darwin goes on to ponder how absurd it must have seemed to imagine that the Earth revolved around the sun, rather than the other way around, when this suggestion was first made. Darwin continues:

> Reason tells me that if numerous gradations from a simple and imperfect eye to one complex and perfect can be shown to exist, each grade being useful to its possessor . . . then the difficulty of believing that a perfect and complex eye could be formed by natural selection, though insuperable by our imagination, should not be considered as subversive of the theory.

I would love to travel back in time and let Darwin know about all the scientific discoveries which have added to our understanding of evolution over the twentieth and into the twenty-first century. I'd tell him about all the incredible fossils which have shown how amphibians grew limbs, how dinosaurs grew feathers and how humans evolved. I'd tell him about DNA and genomes. But in particular, I would love to tell him about the recently discovered genetic links between a 'simple and imperfect eye' and a 'complex and perfect eye'.

Our friend the 'living fossil' lancelet has something which has been known for a long time as a 'frontal eye', even though scientists weren't sure that it was actually used to see with. The 'frontal eye' contains cells which might be sensory, close to dark pigmented cells (just as our own retina contains a layer of light-sensitive cells on top of a layer of pigmented cells). But is this arrangement really light-sensitive? Was the lancelet's 'eye' really a relic of something that could have been the precursor to our much more complex, vertebrate eyes? The potential sensory cells in the 'frontal eye' of the lancelet are very simple, unlike the elaborate rods and cones that are in our eyes. Comparative anatomy, even at a cellular level, couldn't seem to provide an answer.

Enter genetics. A team of Czech and German geneticists and developmental biologists set about solving this conundrum, examining the genes which were being expressed in the lancelet's 'frontal eye'. The genes expressed by the pigmented cells in the lancelet were found to be

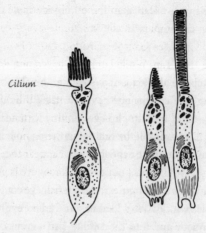

Cilium

A simple ciliary cell in a lancelet's 'frontal eye' (left) and the cone (centre) and rod (right) light-sensitive cells of a human eye

similar to those in the pigmented cells of the vertebrate retina – the 'molecular fingerprint' of these cells was the same. The proposed light-sensitive cells in the lancelet were also expressing similar genes to those expressed in the light-sensitive cells of vertebrate eyes. Importantly, these genes included ones that encoded proteins known to be involved in converting incoming light to an electrical signal in rods and cones. The researchers also tracked long nerve fibres from these sensory cells all the way back to the lancelet's brain. Here at last was proof of a very basic eye in a primitive animal which was related to vertebrates, casting light on how our complex eyes could have started off as something much more simple.

For a free-swimming organism like an early chordate, the possession of a light-sensitive organ up at the front end must have proved advantageous, and as vertebrates diversified, having decent eyes and circuitry in the brain which could receive and interpret images would have helped some to avoid predators and others to find prey.

But there's still a massive leap from something as simple as the 'frontal eye' to something as complex as the vertebrate eye – complete with retina, lens, iris and muscles to move it around. Other living jawless fish may help to bridge this gap. While lampreys have complex eyes much like any other vertebrate, their close cousins the slimy hagfish have very crude eyes indeed. The cone-shaped eyes of hagfish lie underneath a patch of translucent skin, and they possess a retina with nerve fibres that connect to the hypothalamus of the brain. But there's no lens, no iris, no cornea, no eye muscles, and the hagfish doesn't appear to use its eyes for 'seeing' – it's effectively blind; its poor excuse for an eye is probably used to set circadian rhythms. Nevertheless, the hagfish eye could represent a step, one of Darwin's 'gradations', between the basic 'eye' of a lancelet and our eyes. The problem is that, based on current evidence, it's very hard to know if hagfish really are similar to the earliest vertebrates, or whether they were once more lamprey-like, and have degenerated in several ways. That means we can't rely on hagfish to show us possible intermediates between simple and complex eyes.

So we're still left with this leap of faith. Unless, that is, we stop looking just at adult anatomy and consider how eyes develop. The eyes of larval lampreys are very much like those of the hagfish. When the lamprey metamorphoses into its adult form, the eye gets bigger and more complex: cells in the retina become more elaborate, a lens forms, and each eye pops out at the surface and becomes moveable.

We can also get an idea of how eyes might have evolved by looking at how they form in our own embryos. Although we don't run through our evolutionary history in our embryological development, as Haeckel suggested, we can certainly see 'echoes' of our deep past in early embryos. Evolution doesn't make new organisms from scratch, it tinkers with what it already has. In the developing embryo, new features tend to be 'tacked on'.

You're back in the womb, 22 days after your dad's lucky sperm reached the egg. Neurulation has provided you with the basis of your central nervous system, and just as the front end of your neural tube is sealing up, two little protrusions push out from the sides of it, like a snail pushing its eyes out on stalks. Just a few days later, these protrusions have pushed out so far that they've hit the surface layer of your tiny body: the ectoderm. If you had stopped right there you'd have ended up with something that looked like a hagfish's or a larval lamprey's eye. But you keep developing.

Now there's a change at the surface: the ectoderm thickens up, forming a small, round disc, a placode. The optic vesicle protruding from your forebrain begins to collapse in on itself, forming a cup. At the same time, the placode on the surface starts to sink inwards, creating a pit and then pinching off entirely: a bubble of ectoderm trapped beneath the surface of your body. By seven weeks, the spaces inside the cup and the bubble are disappearing. The cup will form the retina (with an inner layer where light-sensitive rod and cone cells develop, and an outer pigmented layer) and the iris of your eye, while the bubble will become the lens. Connective tissue around the optic cup forms a double coat for the developing eyeball, with an inner layer full of blood vessels, and a tough outer layer which will form the 'white' of your eye, as well as muscles which will move the eyeball. Connective tissue inside the optic cup will become the clear jelly of the vitreous humour inside your eyeball. The optic stalk, connecting the original optic cup to the forebrain, becomes the optic nerve. Your eyelids form as folds of ectoderm, fusing over the developing eye and only becoming separate again around twenty weeks after conception, about halfway through your time in the womb.

Going way back, to around the time of the Cambrian explosion, we can imagine a similar series of events or gradations, each 'useful to its possessor', happening over a few tens of millions of years of evolutionary time. Pre-vertebrate ancestors may have started out with a single patch of light-sensitive cells lining part of the brain, similar to the lancelet's 'frontal eye'. As the brain became encased in a skull, this single patch became two, each growing larger and ballooning out to the side. The skin overlying the light-sensitive brain-bulges became transparent. Eyes like this – similar to those of the hagfish – could sense light versus dark, but could not form images. In descendants of this simple-eyed creature, the patch of transparent surface tissue sank inwards, to form a lens. Eye muscles appeared and the retina became more complex. Duplications of genes for light-sensitive proteins (opsins) produced an array of four or five different cone-like cells, each responding to a different wavelength of light. Then rod cells evolved, and the wiring of the retina became more complex too – just as in modern lampreys' eyes. Much later, as some descendants of jawed fish hauled themselves out onto land, eyelids developed to protect the eyes, and the lens changed shape to cope with the way light was bent as it passed from air into the eye.

It's clear, from looking at modern lampreys, that the eyes of early vertebrates were not only complex and image-forming, but that they saw the world in colour. The southern hemisphere lamprey, *Geotria australis*, has five different cone-like light receptors, each with its own type of opsin. Most fish (of the jawed variety) have at least four types of cone, including one which responds to ultraviolet light. The majority of reptiles and birds also have four cones. But most mammals have just two different types of cone, and a relatively small number of cones in their retinas. Among all mammals, it is only an exclusive clique within primates which has developed colour vision again, with three types of cone instead of two.

So, from early vertebrates to us, the story twists and turns: complex eyes have been around a long time, and colour vision is ancient; it is lost and then regained. The colour vision which our fishy forebears enjoyed was lost in our mammal ancestors, with the loss of two of four light-sensitive opsins. To paraphrase Oscar Wilde, to lose one opsin may be regarded as a misfortune, to lose two looks like carelessness. But for any animal that operates largely under cover of darkness, keeping hold of four different colour receptors is unnecessary, or, to put it another way, losing some is unlikely to be detrimental. It is very likely that our earliest furry ancestors were almost exclusively nocturnal.

But why should a select group of primates, among all other mammals, have regained colour vision? Most mammals have two types of cone, one responding to long wavelength light and the other to short wavelengths. In terms of our own colour perception, this makes most mammals red/green colour-blind. Among primates, lemurs, lorises and most New World monkeys are also colour-blind in this way. But, some 30 million years ago, in a monkey which was the ancestor of today's Old World monkeys, apes and us, the gene for the long wavelength opsin became duplicated. One copy of the gene became subtly altered, introducing the possibility of a new cone that would respond to middle wavelengths, and this new cone must have proved advantageous to have been picked up by natural selection.

At the same time as Old World monkeys were diversifying, so were the trees in the tropical rain forests they inhabited. Many monkeys exist on a diet of leaves, so it could be that colour vision helped in spotting younger, more tender leaves. That could have provided enough of a survival advantage for a gene encoding a new cone to spread through monkey populations. This idea – that colour vision evolved to help find food – is not a new one; it was put forward by the nineteenth-century biologist Grant Allen, in a book called *The Colour Sense*, which was published in 1879. After colour vision had evolved, monkeys and apes could take advantage of this new ability, developing forms of colour signalling. Ovulating female baboons and chimpanzees advertise their fertility with brilliantly red bottoms: a signal that would be lost on mammals with a lesser sense of colour.

While colour blindness is extremely rare among monkeys, affecting just 1 in 250 male crab-eating macaques, for instance, it is much more common in Caucasian men, with around 1 in 12 affected. The basis of this colour blindness seems to be that humans have so many copies of the mid-wavelength opsin gene that sometimes these get mixed up with the long-wavelength gene, creating a hybrid opsin which is no longer useful for discerning between red and green. If being able to differentiate between red and green was advantageous for a primate, making it easier to find ripe fruit and tender leaves, it's reasonable to assume that losing this ability would be a disadvantage and would be weeded out by natural selection. Colour blindness may be much more of a disadvantage in monkeys, who don't tend to share food as much as humans do. Colour-blind humans may be less disadvantaged simply because they don't rely solely on food they've collected themselves. So it might be that colour blindness is more prevalent in humans because these defects haven't been weeded out by natural selection as assiduously as they may have been among more selfish monkeys.

While exploring how colour vision evolved in primates, we have been dealing with the mechanics of seeing at a cellular and molecular level. But there are also important characteristics of primate vision which are understandable at a macroscopic level – in other words, at a scale we can appreciate with our own eyes, without relying on microscopes or chemical analysis of proteins.

Hold any primate skull in your hand so that it faces you, and you're staring right down the eye sockets. It seems obvious and ordinary, but you wouldn't get that view with every mammal skull. If you were to look at the front of a mouse, rabbit, sheep, horse, cow or deer skull, you'd just be able to spot the bone around the eye-sockets, sticking out a little at the sides. This is a useful eye position for many mammals: with eyes on the sides of their heads, they get almost 360-degree vision. This is very useful if you're a herbivore and continually on the look-out for predators. In carnivores, such as cats (both big and small), the eyes are positioned so that each eye is gazing out forwards rather than sideways. The visual field of each eye overlaps with the other. This might seem like a waste of space, but it allows the brain to do something ingenious: by comparing slightly different images of the same object, seen simultaneously by both eyes, your brain can work out the distance to that object. You do this neat triangulation without even thinking about it (at least, consciously), but shut one eye and you're left guessing about distances. Most primates have front-mounted eyes with a broad area of overlap between visual fields. Some researchers have suggested that stereoscopic vision evolved in our primate ancestors to help with judging distances – useful if you're an animal leaping from branch to branch, high above the forest floor. But actually primate eyes are rather too close together for this kind of depth perception to be the *raison d'etre* for their stereoscopic vision. Close-set eyes help to judge distances when you're very close to an object, which is handy for an animal that eats insects. And this, it turns out, seems to be what early primates were up to; they probably dined on fruit and nectar, but their main course would have been insects. Depending on a more leafy and then fruity diet came along later, when monkeys and then apes evolved.

Stereoscopic vision lets us see the world in 3D. While you may regret not having eyes in the back – or at least in the sides – of your head, evolution has generously endowed you with very mobile eyes, and a mobile neck which allows you to look back over your shoulder.

There is something quite peculiar about the mobile eyes of humans compared with other primates, and in fact with the eyes of other animals very generally. It is very easy to see which way a human is looking, because of the large amount of the sclera (the 'white' of the eye) that's visible, and the contrast between this and the iris. In most mammals, the sclera isn't white, it's dark, making it difficult to see where the sclera ends and, consequently, making it very difficult to see where the animal is looking. The whiteness of the human sclera may be more than just a random oddity; it's possible that this contrast between sclera and iris has evolved specifically to help us see where someone else is looking – a form of non-verbal communication. We're a very social species, and we're often to be found doing things together, cooperating in all sorts of tasks. Being able to quickly and easily see what someone else is looking at is useful and could have been why our eyes have evolved to be so unusual, in this way, compared with other mammals. But when we look at other apes, it's clear that having a white sclera isn't a uniquely human feature. Some chimpanzees and gorillas also have white sclerae, and gorillas have been shown to pay attention to the direction of eye gaze in other gorillas. But in humans, our eye shape and almost universal whiteness of the sclera mean that it's always easy to see where someone else is looking.

As very visual, social and communicative animals, we pay a lot of attention to other people's eyes. Cicero is quoted as saying '*Ut imago est animi voltus sic indices oculi*', which translates as 'the face is an image of the mind, and the eyes are its interpreter'. Cicero was renowned for his eloquence and wordsmithing, so I think we must be losing something in translation here. Cicero's thought is more poetically expressed in the English proverb: 'The eyes are the window to the soul'. By looking at other people's eyes, we get some idea of what they're thinking about. This goes way beyond following the direction of gaze and working out what another person is looking at.

We also use eye-gaze patterns to make inferences about whether someone is being aggressive, dominant or submissive, about their intelligence, competence and understanding, and about physical attraction. Attention to eye gaze is something we're born with: newborn babies are sensitive to a change in direction of adult eye gaze; by six months old, they will follow the direction of another's gaze when this is indicated by both head position and eyes. By eighteen months, infants will follow eye gaze without the additional cue of head position. In contrast, while adult chimpanzees will follow the direction of gaze of a human, based on head direction, they don't follow the eyes alone. At between two and three years of age, human children begin to use eye gaze to 'mind read' – to work out what someone else is thinking. If you're a member of a social species in which individuals often work cooperatively, this is a helpful skill to master. But it's also very useful if someone is trying to deceive you. For instance, if someone claims ignorance about the whereabouts of a hidden object, they may unintentionally glance at the hiding place. Human adults are very good at picking up on such cues, but can children spot the deceit as easily? Research shows (and parents know!) that children as young as three both understand and practise deception. But do they use eye gaze as a clue, like adults?

Psychologists have used a simple experiment to test children's ability to 'see through' such deception. Children of various ages watch a video where an actor hides a small toy under one of three plastic cups, and the screen goes blank at the crucial moment when the actor hides the toy. In one set-up, the children are told in advance that the actor is tricky, 'she doesn't want you to find the toy'. When the actor reappears on screen, with her three cups and the toy nowhere to be seen, she tells the children that she doesn't know where the toy is, but she glances down at one of the cups. The children have to guess where the toy is hidden. In a second, deceptive, set-up, the actor lies about which cup the toy was under, while glancing at the cup that actually conceals the toy. In another, truthful version of the experiment, the children are told that the actor isn't going to try to trick them. This time, when the actor reappears with the three cups, she says she knows where the toy is and looks at the cup. While three-year-olds have no problem choosing the right cup in the truthful set-up, using eye gaze only as a cue, they're fairly useless at choosing the right cup when the actor is being deceptive, even though they've been warned the actor would be 'tricky'. But four- and five-year-olds find the deceptive set-ups much easier, choosing the cup that the actor looked at even when she claimed ignorance about the whereabouts of the toy, or lied about the toy being under another cup.

These seem like such easy tasks to adults – perhaps we are just a lot more jaded and less trusting? – but just think about what the deceptive task requires. In order to choose the right cup, you have to appreciate that the actor is intending to trick you. Then you have to work out that some information – the words that are spoken – is intended to deceive, whereas other information – the eye gaze – may reveal the truth. The four- and five-year-olds seem to have learned that 'actions speak louder than words'. For three-year-olds, even when they've been told someone is going to trick them, it seems to be very difficult for them to believe that what someone *says* could be unreliable. Four-year-olds are already tuned in to clues given away by someone's eyes. Although our ability to detect deception, and to deceive, continues to improve throughout childhood, even preschool children are able to interpret eye gaze as an important, non-verbal cue.

As well as allowing us to make inferences about the mental states of others, the eyes are also, in a much more literal sense, windows to the brain. Evolving in our ancient chordate ancestors and developing in the human embryo, as outgrowths of the forebrain, the eyes remain connected to the brain by their 'stalks': the optic nerves. If you look into someone's eyes with an ophthalmoscope, and focus on the retina at the back of the eye, you can see where the optic nerve leaves. The optic nerve fibres – about a million of them – bringing information from the light receptor cells right across the retina, converge on a point where they leave the retina and pass backwards to form the optic nerve. The site where they converge is called the optic disc, and it's easily visible with an ophthalmoscope. The optic disc is usually less than 4mm in diameter and it appears bright orange against the deeper orangey-red of the rest of the retina.

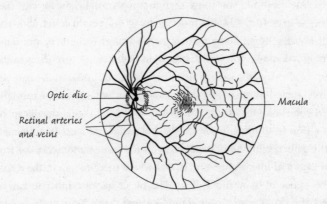

Optic disc

Retinal arteries
and veins

Macula

The retina as seen with an opthalmoscope

There's no room for any receptor cells where the nerve fibres leave the retina, at the optic disc. This is an example of one of the flaws in our bodies which show that we've been 'designed' by a process which has no foresight, no sense of direction. There is no designer: the structure and the function of your body are the products of evolution. If you were to design an eye from scratch, you'd probably put your receptor cells right at the surface of the retina then wire them up so that the nerve fibres were attached to the back of the receptor cells. That way, you wouldn't have to have an optic disc on your retina. You wouldn't have this receptor-less patch, this 'blind spot'.

You may not be aware of the blind spot in each of your eyes, in fact, you shouldn't be. That's because your brain does a good job at filling in the gap – extrapolating from what it sees around the edges of the blind spot. But you can find your blind spot with an easy test that you can do right now. Cover your left eye with your left hand – no peeping. Stare straight ahead with your right eye. Now stretch your right hand out as far it will go, then point up upwards with your index finger. If you're in the right position, you can clearly see the fingernail on your right index finger. Now start to move your pointed index finger around, keeping your arm nice and straight. Move it from side to side and slightly up and down. All the time, you must keep looking straight ahead, you must resist the urge to follow the movements of your right hand with your right eye. It's worth being patient, because something very weird will eventually happen: the tip of your right index finger *will disappear*. It will probably do this when your arm is held out horizontally and very slightly angled outward at the shoulder. It's quite a strange experience when you find your blind spot for the first time; you suddenly realise that vision is so much more than just a passive process. Your brain is working incredibly hard, processing all the information coming from your eyes and making sense of it, and where there's a gap in the data, it makes sense of it by filling it in. The bit of background you can see where the tip of your finger should be *isn't* there. Your brain has just made it up: it's a miniature, visual hallucination. This is fascinating, and now you know your blind spot exists. But if the eye was wired up in a more sensible way to start with, your brain wouldn't have to fill in the gap.

The optic disc may represent one in the eye for any idea of 'intelligent design', but doctors have also found a use for it. Looking at the optic disc can provide a useful clue as to what's going on inside the skull itself.

The outer layers of the eyeball are continuous with a sheath which envelopes the optic nerve, which itself is continuous with the membranes, or meninges, lining the brain. The space between the inner two layers of the meninges contains cerebrospinal fluid (CSF). If there is unusually high pressure inside the skull, the CSF will be pushed out into the sheath of the optic nerve, compressing both the nerve and the central retinal vein which travels inside it. This pressure can be transmitted all the way to the eye itself, causing the nerve fibres in the optic disc to bulge. A swollen optic disc, also known as papilloedema, is a danger sign, signalling a rise in intracranial pressure, which could be caused by inflammation or a tumour in the brain, or a blockage to the flow of CSF. Papilloedema reminds us of the intimate connection between eye and brain, a link that grows out of the embryological development of the eye and takes us back to the origin of eyes in our very ancient ancestors.

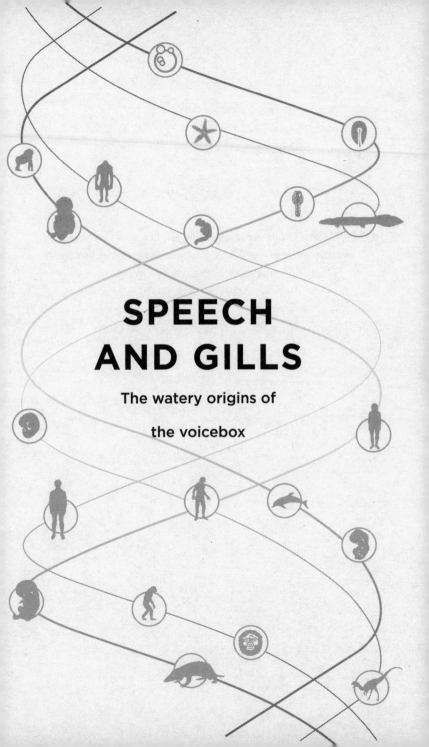

SPEECH
AND GILLS

The watery origins of

the voicebox

'People have to talk about something just to keep their voiceboxes in working order so they'll have good voiceboxes in case there's ever anything really meaningful to say.

KURT VONNEGUT

A U-SHAPED BONE AND BUTTERFLY-SHAPED
CARTILAGE

I n the 1920s, British geologists looking for sources of good-quality stone for the construction of a new port at Haifa, in what was then British Mandate Palestine (now Israel), uncovered some interesting prehistoric stone tools on the western slopes of Mount Carmel. Archaeologists pricked up their ears. Between 1929 and 1934, a team led by the archaeologist Dorothy Garrod excavated a series of caves on Mount Carmel and made some astonishing discoveries. In a cave called Skhul, the archaeologists found thousands of stone tools and ten human burials. It was clear even then that the site was very ancient indeed, going right back to the Palaeolithic. But modern dating techniques have pushed the date back further than Dorothy Garrod may have imagined possible – right back to between 100 and 130,000 years ago. Those skeletons are important evidence: of the anatomy of early modern humans (*Homo sapiens*) and of an expansion of modern humans out of Africa – but that's another story. Other caves on Mount Carmel investigated by Garrod's team, including Kebara Cave, bore traces of human presence in the form of stone tools.

In 1982, archaeologists returned to Kebara to finish what Garrod had started. The team was led by Ofer Bar-Yosef from Harvard University, who was joined by colleagues from France and Israel. Over several seasons of excavation, Kebara yielded its secrets: more stone tools; animal bones from a variety of species, including mountain gazelle, antelope, various deer, wild horse, rhinoceros, hyaena and fox; and the remains of ancient hearths. Then, in the 1983 season, the archaeologists uncovered a human burial, but this time it wasn't a *modern* human, it was a Neanderthal, who had been buried in the cave around 60,000 years ago. Although missing its skull and most of its leg bones, this is the most complete Neanderthal skeleton yet found. Just under the arch of its mandible, the archaeologists found a small, thin, U-shaped bone: the hyoid bone.

This bone is part of your skeleton too; it's tucked away at the top of your neck, just below the mandible. You may be able to feel it, if you make your thumb and index finger on one hand into a 'C-shape' and pinch your neck a couple of centimetres underneath your jawbone. You should be able to detect something quite hard in there, and even wobble it from side to side a little. (Don't spend too much time doing this, or press too hard, as it becomes rather painful.)

The hyoid bone is an important anchor for various muscles and ligaments in the mouth and the neck. The muscles forming the floor of the mouth attach from the inner surface of the mandible at the front to the hyoid bone at the back. A pair of tongue muscles, called hyoglossus, take their origin from the hyoid bone and fan up into the body of the tongue. The stylohyoid muscle is a slender muscle which attaches, together with a fibrous ligament of the same name, from the styloid process (a thin spike sticking down from the temporal bone, under the skull) down onto the hyoid bone. Further down in the neck, the voicebox or larynx is suspended below the hyoid bone, by a sheet of fibrous tissue and some muscles.

The larynx is an extraordinarily complex bit of anatomy. It's only small, about 3cm in diameter and 4cm long, but it comprises nine individual cartilages strung together by ligaments and membranes, and there are movable parts with tiny muscles attached to them. The largest cartilages of the larynx are the thyroid and the cricoid cartilage. 'Cricoid' means 'ring-shaped' in ancient Greek, and the cartilage does look a bit like a signet ring. The thyroid cartilage forms the Adam's apple in your neck, which is much more prominent in a man compared with a woman. The butterfly-shaped thyroid cartilage sits on the ring-shaped cricoid cartilage, with small joints allowing it to rock slightly.

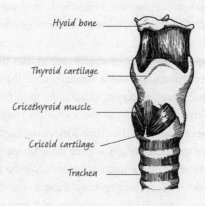

Hyoid bone

Thyroid cartilage

Cricothyroid muscle

Cricoid cartilage

Trachea

The human larynx

On the inside of the larynx, a pair of cord-like, elastic ligaments stretch between the inner surface of the thyroid cartilage at the front, back to two tiny, pyramid-shaped cartilages which sit on the cricoid. (At any rate, *I* think those tiny cartilages look like tiny tall pyramids, but the person who named them apparently thought they looked more like ladles: they are the arytenoid – 'ladle-shaped' – cartilages.) Those ligaments attaching to the arytenoid cartilages are the vocal cords. Muscles attaching to the little pyramid-shaped cartilages, which anchor the cords at the back, mean that the cords can be brought together to close off the opening into the rest of the airway below. This happens automatically when you're swallowing, to help prevent the lump of food from going down into your trachea and lungs. At the same time, the leaf-shaped epiglottis (another laryngeal cartilage) also flaps down over the opening of the larynx to protect it. Apposing the vocal cords to close off the airway is also something that happens when you brace yourself to lift a heavy object (you subconsciously 'know' that you need to hold air in your lungs to help stiffen your trunk), or when you're trying to squeeze out a poo or, indeed, a baby. Sometimes, in such situations, it's hard to stop air squeezing out between the vocal cords, and you might find yourself grunting. (Don't worry, it's entirely normal – and now you know why.)

This role of the larynx – to act as a valve sealing off the trachea and lungs – is its original function in evolutionary terms. But of course it's also the organ of vocalisation: the place where the sound of speech originates.

When you speak, the arytenoid cartilages move the vocal cords close together and pull them tight. When a jet of air is forced past them, it makes them vibrate, creating a sound – very much like the way in which the reed creates sound in a woodwind instrument. The pitch of your voice depends on the length of your vocal cords. In women they're around 1.25–1.75cm long, whereas in men they're between 1.75cm and 2.5cm long. This means that the male larynx is deeper from front to back – it sticks out more in the neck, forming a prominent 'Adam's apple'.

Of course, you can also change the pitch of your voice. You can create a higher note by stretching your vocal cords. This is one of those things you know how to do naturally, without having to think about it. It happens when the cricothyroid muscle at the front of the larynx contracts. This muscle pulls the cricoid up towards the thyroid cartilage. You can feel it doing this in your own neck: find your Adam's apple, then gently move your fingertips down until you feel a slight gap. This is the gap between the thyroid and the cricoid cartilages. Keep your finger there and now make a small, high-pitched 'eek' sound. There. You should have felt the gap disappear as the front of the cricoid was pulled up towards the thyroid cartilage: that was your cricothyroid muscle in action. Because of the way the cricoid is hinged to the thyroid cartilage, it acts like an upside-down see-saw. When the cricothyroid muscle pulls it up at the front, the back of the cricoid, along with the arytenoid cartilages, moves down. This pulls the vocal cords tight, so when you push a puff of air past them, the sound comes out high-pitched.

If speech is such an important human characteristic, we need to ask just how special is the human larynx? It's tempting to imagine that it must be

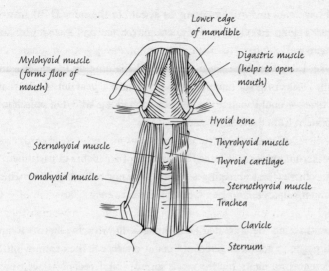

Many muscles attach to the hyoid bone

vastly more complicated than the larynges of other mammals, but in fact it's not. What we have is a pretty standard mammalian larynx, with no added bells and whistles. And perhaps that's not surprising, as it's really the anatomy above your larynx that moulds the raw sound emanating from the voicebox into speech. Although a spoken syllable may start with the sound of the vibrating vocal cords, when we hear someone speaking, we're hearing a lot more than just that buzz. The column of air above the cords changes shape as we speak, and the main factor controlling the shape of the airway and the sound that's produced is your tongue. You can create a burst of acoustic energy by pressing your tongue up against your palate and releasing it, to make a 'd' or 't' sound, or leaving a small gap to force through a 'sss'. Your lips also get involved with making consonants, stopping and releasing the flow of air to make 'b', 'm', 'p', while your lips and teeth come together to create 'f' and 'v'. By changing the shape of your tongue while air flows freely past it, you mould vowel sounds.

I have seen my vocal anatomy in action. In the same fMRI scanner where I glimpsed evidence of my own mirror neurons, I saw my lips and tongue working hard to create a stream of vowels and consonants as I spoke. I was filming a piece about vocal anatomy, and I spoke words while I was inside the scanner that I would later repeat outside, in a cafe, so that we could match the film of me talking with the animations produced from the fMRI. I said:

Air from my lungs is forced between my vocal cords causing them to vibrate. The sound passes upwards and is moulded by my tongue and my lips, emerging as speech . . . ahh . . . eee . . . ooo.

I loved seeing the animated fMRI images, showing my larynx, tongue and palate in action as I spoke. I could clearly see the anatomy of the different consonants, but the vowel sounds – ahh, eee, ooo – were most interesting as I wasn't really aware of the contortions that my tongue was throwing itself into, inside my mouth. In 'ahh', my tongue was bunched up at the back of my mouth; in 'eee', it was pushed up in the middle, lying quite close to my palate; in 'ooo' it became a strange shape, hunched up, with a gap at the front.

In making intelligible speech, our tongues are incredibly important – and they *are* different from those of other mammals, including our closest primate relatives, chimpanzees. Partly because the very short 'snout' that we humans have, our tongues are not long and flat like those of most other mammals, but rounded, and highly mobile. But tongues, like larynges, don't fossilise. So if we want to know when our ancestors developed large, mobile tongues, we're left looking for clues in the skeleton which might give us some insight into soft tissue anatomy – and it's tricky. Even if we can't look directly at the muscle of the tongue in our ancestors, it's possible that there might be a clue about tongue muscle in the nerve supplying it: the hypoglossal nerve. Named nerves in the body tend to be ones that are big enough to see with the naked eye, and they are bundles of often hundreds of thousands of individual nerve fibres.

Each of the fibres in a motor nerve like the hypoglossal nerve will eventually reach the muscle, and supply a group of muscle fibres called a 'motor unit'. So the size of the hypoglossal nerve, which depends on the number of nerve fibres inside it, should also give an indication of the number of motor units in the tongue. A greater number of motor units may make for a more finely controlled tongue – something that sounds like it would be very useful for speech.

But nerves are soft tissue too, and they don't fossilise either. However, the hypoglossal nerve is one of the 'cranial nerves' – nerves which emerge from the brain itself and most of which have to escape out of the skull to supply structures on the outside. (There are twelve pairs of cranial nerves, including some we've already met: the optic, olfactory and vestibulocochlear nerves.) The hypoglossal nerve exits the skull through its very own channel, the hypoglossal canal, in the skull base, close to the large foramen magnum where the brainstem emerges from the skull to become the spinal cord.

In the late 1990s, a group of anatomists at Duke University, in Durham, USA, went about measuring the size of hypoglossal canals in the skulls of humans, chimpanzees and gorillas, and in a few fossil hominins. It seemed the scientists were on to something. On average, the size of the hypoglossal canal in humans was around twice that in chimpanzees and a third bigger than in gorillas. In australopithecines, early hominins who lived in Africa around two to four million years ago, the canal was small, similar to chimps and gorillas. When the team looked at later hominins, including a skull of *Homo heidelbergensis* (the common ancestor of Neanderthals and modern humans) and a Neanderthal, the hole was bigger: human-sized. If a large hypoglossal canal means a larger nerve supplying the tongue and a human-like capacity for speech, then it looked like this important aspect of human-ness was already in place by at least 400,000 years ago, in our proximate ancestors, and then in our cousins, the Neanderthals.

It's a great story. Unfortunately, just a year after those results were published, a second team of American researchers reported findings which cast doubt on the very neat hypoglossal hypothesis of speech evolution. They measured hypoglossal canals in 104 modern human skulls and 75 other primates. In this larger sample, they found a much bigger range of human hypoglossal canal sizes, which overlapped with the size of the canal in other apes, and in ancient hominins. In fact, the primates with the largest hypoglossal canals turned out to be monkeys and strepsirhine primates (lemurs and lorises), and no one is suggesting that they can talk! A further nail in the coffin of the hypoglossal hypothesis emerged from a dissection study which showed that the size of the hypoglossal canal wasn't a good indicator of how big the nerve actually was.

So much for the hypoglossal canal, then. But what about the hyoid bone? Could that help to shed some light on the question of the origins of speech? The discoverers of the Neanderthal hyoid, in the Kebara Cave in Israel, certainly seemed to think so. In their report on that hyoid bone, they argued that the Neanderthal hyoid bone was so similar to a modern human one that Neanderthals must have had a fully developed capacity for speech. Two further hyoid bones, from the famous Sima de los Huesos site in the Sierra de Atapuerca, in Spain, and belonging to archaic Neanderthals, are also very much like ours. There's also one fossil hyoid bone belonging to an earlier hominin species, *Australopithecus afarensis*, and this one is much more chimpanzee-like. So perhaps australopithecines were incapable of speech while Neanderthals would have been able to chat away.

A big claim for such a small bone and perhaps a step too far. The larynx may be suspended from the hyoid bone, but we still don't get any information about the soft tissue – including the cartilages of the larynx – from looking at the hyoid. If I was feeling particularly catty, I'd suggest that trying to recreate all that soft tissue, including the larynx itself, based on just the hyoid bone, is like trying to reconstruct a dress based on a coat hanger.

In fact, we may never be able to settle the question of *when* our ancestors developed the power of speech by looking at anatomy. As we've already seen, the human larynx itself is fairly unremarkable: a bog-standard mammalian piece of kit. But there *is* something odd about your voicebox, compared with other mammals, and that is its position in your neck: it's very low. This means that there's a long vocal tract between the vocal cords and the lips, long enough to allow a broad range of different sounds to be produced. The tongue, soft palate, lips and teeth modify the puffs of air emerging from the larynx into consonants and vowel sounds, as I saw in the MRI animations of my own vocal tract.

The American cognitive scientist Philip Lieberman has spent decades investigating the anatomy, function and evolution of the human vocal tract. He argues that the low position of the adult human larynx is important because of what it means for the position of the tongue. A low larynx and a low hyoid bone mean that the back of the tongue is in such a position that it can affect the dimensions of the pharynx (behind the tongue) as well as the oral cavity (above and in front of the tongue). The right-angle bend between the oral cavity and the pharynx in humans also effectively breaks up the vocal tract into two tubes of equal length, and Lieberman thought that was important to the ability to produce discrete vowel sounds – 'ahh', 'eee', 'ooo'. Being able to produce acoustically distinct vowels may sound like a luxury, but it gives our brains half a chance of decoding what another person actually means when they make these different sounds. Neanderthals can't have had vocal tracts with two tubes of equal length like us, because they have long jaws, making for a very long oral cavity. If they had an equally long pharyngeal tube, above the vocal cords, this would have pushed their larynges right down into their chests, which is an anatomical impossibility. Would this have made their vowels indistinguishable from each other, and their speech unintelligible? Lieberman thought so, but other scientists disagree: ten-year-old children have a similarly proportioned vocal tract and they are, on the whole, fairly easy to understand.

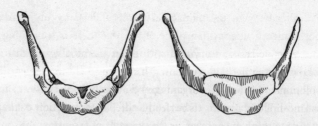

A modern human hyoid bone (left) and the Neanderthal hyoid bone from Kebara

So what would a Neanderthal have sounded like? Acoustic scientist Anna Barney, together with anatomist Sandra Martelli, and their colleagues at the University of Southampton and UCL, recently set about trying to find this out (as well as trying to answer more serious questions about whether Neanderthals could produce acoustically different vowels). The team made virtual reconstructions of vocal tracts based on CT scans of living people and of fossil Neanderthals, and then used a computer model of speech production to see how different the Neanderthal would sound compared with a modern human.

In the computer model, the predicted position of the Neanderthal hyoid bone placed it just below the jaw, just as it is in your neck, but the distance from the hyoid bone to the front of the mouth was longer than in a modern human because Neanderthals have longer jaws. The distance between the hyoid bone and the cervical spine – the space in which the upper pharynx sits – was also longer in the Neanderthal. This means that the Neanderthal tongue couldn't push back into the pharynx as much as it does in us, and that has an effect on vowel sounds, which was demonstrated by the acoustic modelling. While the 'eee' and 'ooo' vowels were similar to those in modern humans, the virtual Neanderthal vocal tract could not produce a completely modern-human-like 'ahhh' – the vowel produced when the tongue bunches up at the back, modifying the shape of the upper pharynx behind it, exactly as I'd seen in my MRI.

LARYNGEAL MYTHS

The descended human larynx and its role in speech is a great illustration of how fascinating and frustrating it can be to try to reconstruct the evolutionary history of an organism and its parts. It's very easy to assume that most of our anatomy is perfectly suited to its function. Perhaps we still like to think of ourselves as the pinnacle of evolution, the top rung on a *scala naturae*, rather than a twig which has managed to survive in the part of the tree of life which has otherwise been rather heavily pruned.

But if you think about it for a moment, it can't be true that every single anatomical element in your body is perfectly suited to its function, and most bits of anatomy subserve more than one function. For some structures one function may be particularly important, and, through natural selection, may push the form a long way in a certain direction, but the need to fulfil other functions is likely to exert a tension. There are inevitably going to be trade-offs. This means that the final form of a particular structure is likely to represent a design compromise, whichever function you look at.

It's also tempting to see every bit of anatomy as an adaptation, as something that natural selection has worked on, to fit its current form to its present function. We assume that the human larynx has descended in the neck to make vocalisation better, perhaps clearer. The adult position of the human larynx, low down in the neck, means that the vocal tract comprises two tubes of equal length – something which seems necessary for producing clearly distinct vowel sounds.

The traditional story about the evolution of the uniquely descended human larynx also points out that a low larynx is a potentially life-threatening hazard: it makes choking much more likely. The benefits of speech must have outweighed the higher risk of choking, in evolutionary terms. In other mammals, the larynx is higher and the tip of the epiglottis reaches high up, above the soft palate. This means that the pharyngeal airway is kept separate from the food-way. In humans, the position of the larynx is much lower in the neck, and the tip of the epiglottis is far below the soft palate. As a lump of food is swallowed, it has to pass over the top of the larynx, interrupting the airway and potentially slipping down into it.

I've always been suspicious about this story. Firstly, my dog frequently almost chokes on overly large mouthfuls of food (but thankfully manages to remove the obstruction to his airway by coughing up offending bits of food onto the floor). Secondly, an animal's epiglottis may reach up to its soft palate when it's *not* swallowing, but when it does swallow, the epiglottis will flap down, losing its connection with the soft palate, and the ball of food has to pass over the larynx to get down the pharynx, into the oesophagus – just as it does in you or me. Despite the lower position of our larynges, we swallow in much the same way as most mammals.

The evolutionary anthropologist Leslie Aiello doubted the veracity of the traditional story about choking too. When she looked at UK death registers over the last hundred years she found that, firstly, choking was a rare cause of death, and secondly, there didn't seem to be any increased risk of choking in adults with descended larynges, compared with children whose larynges were still high in the neck.

So the argument that selection pressure for a descended larynx for speech must have been strong enough to override the risk of choking to death doesn't stand up to scrutiny. And here's another thing: what if the descended human larynx *isn't* an adaptation for speech at all? What if the larynx happened to be in a low position for an entirely different reason, and that turned out, in the long run, to be quite useful for speech?

Your head is balanced on top of your cervical (neck) spine. A chimpanzee's cervical spine articulates with its skull further back. If you look at the base of a human skull compared with a chimpanzee skull, it's clear that the position of the foramen magnum – the large hole where the brainstem exits the skull before it becomes the spinal cord – is different. In a chimpanzee, the foramen magnum is much closer to the back of the skull than it is in a human. On each side of the foramen magnum (in both species), there are two oval surfaces known as the occipital condyles – this is where the skull articulates with the first vertebra of the spine.

We humans are habitual bipeds; we regularly walk and move around on two legs. Our anatomy reflects our bipedality, and it's not just our legs and pelvis that have had to change to accommodate our commitment to bipedalism, our spine has changed too, developing into a double-S shape. And the reorientation of the attachment of the head to the spine has happened because of the way that we hold our bodies: our heads are balanced on the top of an upright spine. This is why the foramen magnum, and the occipital condyles, are so much further forward in the human skull. The growth of our big brains has also played a role, as the back of the skull, behind the foramen magnum, has bulged out.

But look at the bottoms of these two skulls again. Notice how close the foramen magnum is to the back of the hard palate in the human skull. It's much further away in the chimpanzee skull, which means there's room to stack structures in the neck, behind the jaw. When our

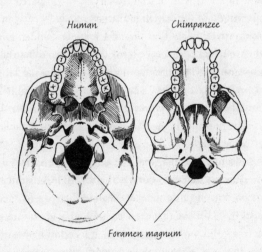

The foramen magnum is positioned relatively further forward in a human skull

ancestors took to regularly walking around on two legs, the larynx had to move down, otherwise it would have been squashed between the tongue and the spine, which would have made both breathing and swallowing impossible. In other words, the low position of the human larynx may be a knock-on effect of walking on two legs, rather than an adaptation for speech. As bipedalism is the earliest 'human' characteristic to appear within our hominin ancestors, this would suggest that the larynx descended very early in human evolution – long before any of our antecedents started to speak.

However, evolutionary anthropologist Dan Lieberman argues that bipedality may not have forced laryngeal descent on its own. He suggests that the protruding jaws of our australopithecine ancestors would have meant that there was still plenty of room for a high larynx. It was only with the appearance of our own genus, *Homo*, with characteristically flatter faces (which may have come about due to changes in diet, perhaps even including cooking to soften food, which led to smaller teeth), that the larynx would have been squeezed out and would have had to descend in the neck.

Pushing the origin of the descended larynx relatively early in human evolution, and providing another reason for its descent, still leaves us guessing when our ancestors developed the crucial human ability of speech. Even if early *Homo* had the necessary anatomical kit in their mouths and necks, it could have been there for hundreds of thousands of years before they used it to produce speech. The important anatomical changes that permitted speech happened inside the skull, in our ancestors' growing brains.

Attempts to recreate Neanderthal vocal tracts have shown that they probably had different voices to ours, perhaps with less distinct vowels. There seems no reason to doubt that Neanderthals could speak (and many archaeologists believe that their technology and culture strongly suggest that they did possess language). It's certainly possible that natural selection has acted to improve modern human speech capabilities, compared with Neanderthals. But we could also have started off with an advantage for other reasons: our even flatter faces and shorter jaws may have serendipitously created a vocal tract with two equal-length tubes and, as a result, clearer vowels.

The relative position of the larynx changes throughout postnatal development and growth. By eight years of age the 'two tubes' have reached that ideal one-to-one ratio, and although everything is still growing for a number of years, that ratio is maintained. At least, it is in girls, even as they pass through puberty. In boys, of course, it's different. As their voices break, their vocal cords grow longer, but their larynges also descend even further in the neck, ruining that perfect one-to-one ratio.

So if there's an adult vocal tract that nears perfection – again, for that important ability of being able to produce clearly different vowels – it's to be found in women. Recent studies of modern human vocal tracts have shown that the female tract produces a larger range of discrete vowels than the male. The continued descent of the larynx in adult men certainly wouldn't have provided any kind of evolutionary advantage linked to improved articulation of speech. But while they may lose a slight edge in clarity, men gain something else from their low larynges: deep voices. Low male voiceboxes could have come about through sexual selection: if women way back in our deep ancestry found men with deeper voices more attractive, perhaps because a deep voice suggests a large body size, then men with low larynges would have had an advantage when it came to passing their genes on.

DEEP-VOICED MALES

It might seem, from all this discussion about the descended larynx, that this is a uniquely human characteristic. But it's not. Male red deer and fallow deer also have low voiceboxes, which get even lower during the rut, when the stags let each other know how big they are by roaring. Male red deer can be very persistent, roaring up to twice a minute, all day. Fighting is costly, and roaring allows males to assess each others' size and fighting ability before settling on that risky course of action.

In autumn, red deer hinds gather and stags come to meet them. Mature stags gather harems of hinds around them, which they're very keen to defend, keeping the hinds near and their competitors at antler's length. They're so keen, in fact, that they almost stop eating in order to do so and thus can lose up to 20 per cent of their body mass.

When a challenging stag approaches a harem-holding stag, and especially if the two stags look fairly equally matched in size, there's often a roaring contest. Only a few of these contests end in violence. Studies of roaring contests which did end in fights have shown that comparing the roaring rates of the two stags is a very good way of predicting which one will win: the winners roar more often.

The position of the larynx in male red or fallow deer is comparable to that in humans, but obviously this has nothing to do with speech. Instead, this is all about roaring – the deeper the better. When the stag roars, his larynx gets even lower in his neck; the strap muscles, which run from the sternum to the thyroid cartilage of the larynx and on to the hyoid bone above, undergo a steroid-boosted expansion in the run-up to the rut. When the stag roars, these muscles drag the larynx even lower in the neck (lower than it ever gets in humans), lengthening the column of air in the pharynx, above the vocal cord, and deepening the stag's voice. In his 'two-tube' vocal tract the pharynx is now twice as long as his oral cavity. In red deer, the larynx is pulled right down as far as the top of the breastbone – in other words, as far as it could possibly get.

Low-frequency sounds travel further than high ones, so it's possible that the deep roar produced by such a low larynx provided stags with an evolutionary advantage because it's far-reaching. But the problem with this hypothesis is that most roars seem to be directed at the ears of other males who are rather too close for comfort, as far as the roarer is concerned. The alternative is that a particularly low voice strongly implies a large body size. By lengthening his vocal tract, the deer is (unknowingly, of course) exaggerating his body size.

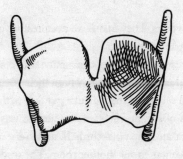

The thyroid cartilage of the larynx

It seems that there are other mammals that also have descended or retractable larynges, suspended by very elastic membranes and ligaments. Unsurprisingly perhaps, these are also animals that roar: lions and tigers, and trumpeting elephants. It may seem that we're going off on a bit of a tangent here, but all of this is relevant to human evolution because it means that there are even more reasons for having a low larynx, beyond being able to speak. It turns out that the human larynx is even lower than it needs to be just to get it out of the way in a creature with a skull resting on an upright spine and a short face. But even if we could work out exactly when the larynx sunk to its present depths, we still couldn't be absolutely sure that this happened as an adaptation for speech. It could have been for size exaggeration, as seems to be the case in other mammals. Because, as far as we know, deer, lions, tigers and elephants don't speak.

Perhaps these other mammals are also showing us why particularly deep voices have evolved in human males. It could be simply because women have preferred bass voices (and possibly still do), a bit like the 'Barry White effect'. A recent study from the University of Aberdeen found that women not only expressed a preference for deep voices, but also remembered things better when they'd been told them by a man with a low voice! But it could also be that men, like deer, have evolved to exaggerate their body size through vocalisation in order to deter sexual competitors. It seems that men are, subconsciously, quite clever about the pitch of their voices. They may have a set 'resting pitch' but they can change that pitch, and they do.

In a study carried out at the University of Pittsburgh, 111 male students took part in a dating game, where their voices were recorded. First, each student was asked to read out a passage, so that the researchers could get an idea of his 'normal' voice. Then he was told that he'd be competing against an unseen (recorded) 'adversary' in another room, for a date with a woman in yet another room. The student volunteer was also recorded having a conversation with his adversary. Finally the student had to fill in a questionnaire which included items about how dominant a person he felt himself to be.

But the really fascinating finding of this study is that participants who believed themselves to be more physically dominant than their adversaries in the other room, lowered their voice. When a student participant thought they were less dominant, they raised their voice to a higher pitch. So this suggests both that men will rate the physical dominance of another individual based just on their voice, and that they will modulate the pitch of their own voices accordingly.

If we think about roaring red deer, this may make more sense. If a male human thinks he's dominant, he'll emphasise that dominance by lowering his voice, intimidating his competitors and avoiding conflict in that way. The message is: he's going to get the girl, and there's no point in fighting him for it. If, however, he thinks he's less dominant, then a higher voice appeases his more dominant competitor: don't bother fighting me, you've won, mate.

Competition for mates involves more than just female preferences for males, it also brings in competition among each sex for access to a mate. In this case, a low voice might help a man to exclude other, apparently less dominant, men from the mating game, and it might make him more attractive to the opposite sex. It's a win-win situation, unless, of course, you're a well-built man with a mellifluous tenor voice who's bucked the trend, or a woman who prefers Justin Timberlake or John Lennon to Jim Morrison or Eddie Vedder.

A VOICE FROM GILLS

Putting our recent human evolution to one side and journeying much further back in time, we can find out where a larynx comes from in the first place. It's something that is only possessed by vertebrates whose ancestors made their way out of water onto land. Fish don't have larynges. But, how did the larynx develop? Bits of anatomy don't generally just appear out of nowhere, there has to be a precursor, *something* that can be modified, duplicated or have a little extra added to it. So where does the larynx come from?

There are clues in the anatomy of adult vertebrates, but the answer only really becomes clear when we start to look at embryos. This takes us back to the origin of vertebrates, and that novel population of nomadic cells which comes from the crest of the fusing neural tube. Neural crest cells end up in a lot of different places – in your skin, in your heart and in your adrenal glands, to name a few destinations. Some neural crest cells find their way into the tissue flanking the pharynx and turn into skeletal, supporting structures there. They make what's called the 'splanchno-cranium', which forms the skeleton of the face and other hard(ish) tissues in the neck. At this point, you need to stop thinking about the skeleton as being made of bones only: it's made of cartilage too – even in an adult. Cartilage lines the mobile joints of our body; it joins the ribs up to the breastbone and it forms the skeleton of the larynx – which we've already met in the form of all those cartilages: thyroid, cricoid, arytenoid, the epiglottis, and some others too small to mention.

In most fish, this splanchnocranium includes bones and cartilages which form and support the jaws and gills.

The diagram overleaf shows the splanchnocranium in an adult shark – including Meckel's cartilage, forming the front of the lower jaw or mandible. Below that, a hyomandibular cartilage attaches the jaws to the otic capsule of the skull (which contains semicircular canals, just as it does in us), and under that, there's a series of cartilage struts which

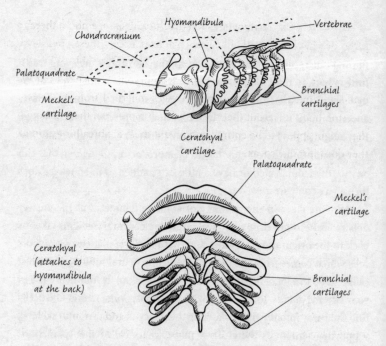

Two views of the jaws and gill arches of a shark, from the left side (top) and from below (bottom)

help to support the gills. Sharks and rays are cartilaginous fish – their entire skeleton is made of cartilage which never turns to bone.

The pattern of the splanchnocranium is very similar in bony fish (which include just about every other fish you can think of that isn't a shark, a ray, a hagfish or a lamprey), but although all these structures start off as cartilage in the fish embryo, they eventually ossify, or turn to bone. As well as becoming bone itself, Meckel's cartilage in the jaw also becomes enveloped in a new layer of bone which develops straight from mesenchyme (that undifferentiated embryonic tissue), without a cartilage stage, just like the flat bones in the vault of your skull. The jaws and hyoid of fish, together with bars of cartilage or bone which support their gills, develop in a series of branchial (gill) bars, along the sides of the embryonic fish's neck.

Here's a remarkable thing: in a five-week-old *human* embryo there's a series of what look exactly like branchial bars, except these ones never develop into gills. More's the pity, as I would love to be able to breathe underwater. So what happens to our own embryonic branchial bars? We could presume that they're merely vestiges inherited from an ancient ancestor, but it turns out that they are much more than that. Although they might appear to be entirely redundant in an air-breathing animal, they don't just disappear.

We last left your developing embryo just as you were making yourself a neural tube, in the fourth week of development. At the same time as the neural groove is zipping up to form the neural tube, the edges of the three-layered sandwich (the trilaminar germ disc) started to curl down. Eventually, those edges will meet, and instead of a flat sandwich, you've now got nested cylinders: a tube inside a tube inside a tube. Having started off at the beginning of week four as a flat sandwich, you now look something more like an animal, with a head and a tail. The ectoderm on the top of the trilaminar-germ-disc sandwich now covers your entire body, not just your back. The lining of your gut comes from endoderm, and this layer is hidden away inside your body, not exposed on your front. In between the ectoderm and endoderm is the jam in the sandwich: mesoderm. Now the embryo is really cooking: the recipe is starting to work, but we're making something much more complicated than Mary Berry or Paul Hollywood could even dream of.

Although it might not feel like it, these three nested cylinders represent the basic structure of your body. The outer tube of ectoderm will form the outer layer of your skin: your epidermis. The inner tube is your gut, from mouth to anus. In between the lining of your guts and your outermost layer of skin, mesoderm turns into everything else, including organs, bones and muscles.

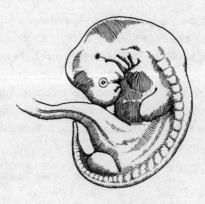

A five-week-old human embryo

The embryo doesn't rest, and as you embark on week five, all sorts of other interesting things are starting to happen to this tiny embryonic you. Up in the neck, those branchial or pharyngeal bars have appeared: five of them on each side. The branchial bars are visible from the outside while their inner surface flanks the pharynx – the upper end of the gut tube – of your developing embryo. The outside of the bar is covered with the same layer that coats the outside of the whole embryo: ectoderm. The inner surface is the lining of the gut tube: endoderm. In the middle is that loose embryonic tissue called mesenchyme, some of it from mesoderm and some from immigrant neural crest cells. This mesenchyme will develop into a cartilage bar and muscles, and each branchial bar also contains an artery and a nerve.

In a fish, the components will all develop into essential parts of gills: the cartilage will support the gill; the artery will bring deoxygenated blood from the heart to the capillaries of the gill, where the blood becomes oxygenated before entering a large artery running just under the spine of the fish: the dorsal aorta. Some of the mesenchyme inside an embryonic fish's branchial bar will become muscles which open and

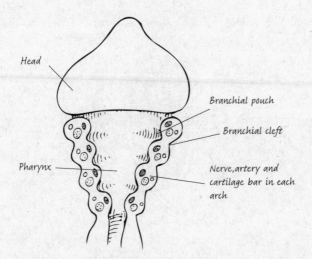

A cross section through the pharynx and branchial arches of a human embryo

close the gills. The nerve running into each bar will supply those muscles, as well as supplying sensation.

You and I have no need for any of this gill apparatus. When our fishy ancestors crawled onto land, they needed a new way of extracting oxygen from their environment, and they swapped gills for lungs. Over vast stretches of evolutionary time, branchial bars and all that they contained became subject to the re-purposing and recycling that is grist to the mill of evolution. Evolution is very good at re-purposing or recycling structures; the evolutionary biologist Stephen Jay Gould used the analogy of turning car tyres into sandals: something that started out as designed for one very particular purpose could be turned into something completely different. In this case, bits of what were once gills have been recycled to make a voicebox.

Tracing the fate of the cartilages in the lowest two branchial bars in a human embryo reveals that they become the cartilages of the larynx. Muscles which would have opened and closed the gills of your fishy

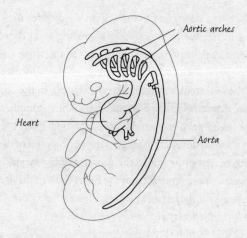

*The branchial arch arteries carry blood from the heart back to two dorsal aortae –
which fuse to form one aorta lower down*

ancestors have become the small muscles of the larynx, including ones which move the vocal cords. The nerve supplying these muscles is an especially quirky bit of anatomy.

THE U-TURN OF THE RECURRENT LARYNGEAL NERVE

The tenth cranial nerve (also called the vagus nerve) emerges from the base of the skull and runs down the neck, giving a branch to the larynx. But this branch doesn't make a beeline for the larynx, far from it; instead it continues descending in the neck, right down into the chest, before doing a U-turn and ascending back up to reach the larynx. It's called the recurrent ('running back') laryngeal nerve.

The journey of this nerve is one of the perplexing oddities of adult human anatomy – and indeed it's even more extraordinary in a giraffe, travelling all the way down and then back up the giraffe's neck. Why on earth would a nerve lose its way so badly, looping right down into the chest before retracing its path, back up to the larynx, when a neat, horizontal branch in the neck would be much more efficient? In fact, that's exactly how the recurrent laryngeal nerve starts off, in the embryo. But there are other structures in the branchial bars which complicate things: arteries in the branchial bars connect the embryonic heart to the aorta.

Your heart starts to develop right up in the neck region of the embryo. When the nerves first grow out to the muscles of the larynx, they pass below the lowest artery running from the heart back to the aorta. Much later, your embryonic heart moves down into your thorax, dragging those arteries – and the laryngeal nerves – down with it. The arteries may get completely re-plumbed, but the nerves are now trapped in your chest, the one on the left looping under the arch of your aorta, and the one on the right running under your right subclavian artery (which becomes the main artery in your right arm).

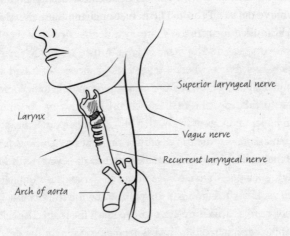

The left recurrent laryngeal nerve branches off the vagus nerve, loops under the arch of the aorta, and travels back up the neck to the larynx

It's a quirk of anatomy that seems unnecessarily complicated – unless you know a bit of embryology, and then it all makes sense. It makes even more sense if you bring in an evolutionary perspective as well. An animal is never designed from scratch, it is just a slight modification on what went before. When it comes to nerves and blood vessels in particular, this can make for some very complicated adult anatomy, because you're looking at a set of electric wires and plumbing that has been tweaked over millions of years. If you buy an old house with tangled and outdated wiring, you have the option of replacing the whole lot. Evolution doesn't have that luxury, bodies cannot be completely rewired. It makes anatomy more complicated, certainly, but it also means that our bodies hold these incredible clues to our own, very ancient, evolutionary past.

JAW JOINTS AND OSSICLES

The other branchial arches also have an extraordinary evolutionary and embryological tale to tell. The first arch contains a very ancient story because, in most fish, it's already a double structure. Looking at the embryological development of jaws in fish, we might imagine an evolutionary scenario where a filter-feeding ancestor-fish had a single first branchial bar, which had to 'break' in the middle to become a pair of jaws. In this way, the first branchial bar ended up forming parts of both upper and lower jaws, with a hinge in the middle: the jaw joint. The cartilage bar in the lower part of the first branchial arch, which contributes to the mandible, is named after its discoverer: it is Meckel's cartilage.

The first branchial arch is also important in the story of a particular group of reptiles which evolved to become mammals. These reptiles underwent some incredible changes, swapping scales for fur, developing milk-producing glands, and eventually, in the ancestor of most living mammals, giving up laying eggs to give birth to live young. In the jaws and ears of early mammals, there were other, perhaps more subtle, but just as important, changes occurring.

Reptiles have just one ossicle linking the eardrum to the cochlea of the inner ear, equivalent to the stapes in your ear. But you, like any other mammal, have two more ossicles in each ear: an incus and a malleus as well as the more ancient stapes. The new three-ossicle system in the middle ear, linking the eardrum to the cochlea, gave mammals an important advantage in being able to hear higher-frequency sounds than their reptile predecessors could – a useful ability if you're creeping around at night, looking for insects to eat, while trying to avoid being eaten by the dominant life-forms on the planet. Mammals evolved when dinosaurs ruled the Earth.

We can trace where those two new ossicles came from, evolutionarily, by looking at their embryological development. They both come from the first branchial arch, in fact, they're part of what becomes the jaw joint in reptiles. Mammals were able to use this old jaw joint to make two new ossicles because they'd 'invented' an entirely new jaw joint to replace the old reptilian one.

However, that still leaves us with an evolutionary conundrum. It seems like an impossible trick to pull off, an amazing sleight of hand: 'Before your very eyes, behold! The old jaw joint is in the ear and a new jaw joint takes its place!' Could the switch really have been so swift? It seems like the evolutionary equivalent of pulling a tablecloth out from under a tea set. It seems biologically unlikely that a new jaw joint could have appeared at *exactly* the same time as the old one was stolen away to become an extra pair of ossicles.

This remained quite a mystery until the discovery of little *Morganucodon*, an early mammal which lived around 205 million years ago. Many of the fossils of this small, shrew-like creature come from Glamorgan in Wales (its name means 'Glamorgan tooth', which sounds like a terrible affliction). *Morganucodon*'s jaw is quite extraordinary: on each side, it articulates with the jaw in *two* places. This mammal had different teeth to its reptile predecessors: *Morganucodon*'s teeth slid past each other in a shearing action. This type of biting applies a twisting force to the jaw, and the addition of a second jaw joint might have helped to resist this. But once a new jaw joint evolved, the original one was freed up to become part of the hearing apparatus. *Morganucodon* is the anatomical bridge between reptiles with their ancient jaw joints and only one ossicle in each ear, and mammals, with a new jaw joint and two new ossicles. Mystery solved, and no one was left jawless in the process of this transformation.

In fact, the trick of stealing bits of jaws for bits of ear was nothing new. Even before mammals evolved, reptiles were already in receipt of such stolen goods. The hyomandibular cartilage braces the lower jaw of sharks against the skull, and the same structure exists – as a bone – in bony fish: the hyomandibula. In land animals, the upper jaws became firmly attached to the skull, meaning that there was no need for extra bracing. The hyomandibula was now surplus to requirement in the jaw, but it stayed attached to the otic capsule of the skull, becoming the stapes.

There's a tiny muscle attached to the stapes in your ear. It is, in fact, the smallest muscle in your body and it's called stapedius. When it contracts, it pulls on the stapes and dampens its vibrations. It's an extremely important safety device: as soon as you start to hear a very loud noise, stapedius contracts to stop large vibrations from being transmitted to the cochlea, where they would damage the delicate hair cells. And while your stapes is equivalent to the hyomandibula which braces the jaw in fish, your tiny stapedius muscle is homologous with a muscle which opens the jaws in fish and reptiles.

Other muscles originating from the second branchial arch in the embryo, just like stapedius, have ended up on our faces. Muscles which operated the flap covering the gills in our fishy ancestors give us our expressive faces, including the ability to smile.

The most important muscle when it comes to a beaming smile is zygomaticus major. This muscle is attached at the top to the cheekbone, or zygoma, and then inserts into the corner of the mouth. It's supplied by branches of the facial nerve, which also supply all the other 'muscles of facial expression', which all develop from the second branchial arch. Look in a mirror and smile; on each side of your face, zygomaticus major is pulling the corners of your mouth upwards and outwards. If you happen to be one of the people who has a split zygomaticus major muscle, where the lower part of it is tethered to the overlying skin, this will create a dimple in your cheek when you smile.

You can widen your smile: other muscles act to raise the whole of the upper lip, and to pull down the lower lip, exposing the teeth. Try smiling even more widely and you're co-opting another muscle, risorius (named after the Latin for 'grin'), which can also help to pull the corners of the mouth further back. But now try using risorius on its own: let your face relax, then pull back the corners of your mouth. This produces an unpleasant grin or grimace: something which looks more threatening or fearful than joyful.

Our close primate relatives have very similar expressive muscles in their faces, derived from the second branchial arch in their embryos, just like ours. All primates have what's called a 'bared teeth' expression, but while many anthropologists in the past suggested that this was essentially similar to a human smile, others said it was closer to a fearful grimace. More recent research has revealed a complex picture. Perhaps unsurprisingly, the 'bared teeth' expression seems to mean different things in different species. In some macaques, it's a sign of submission, but in Gelada baboons and chimpanzees it's more appeasing and helps social bonding – in other words, it really does seem to be the equivalent of a human smile. The 'bared teeth' expression in chimpanzees certainly looks similar to a smile: the corners of the mouth are pulled up and the lips parted (though to an even greater degree than in the human 'complex smile'). Under the skin, chimps have almost identical facial muscles to us, including those zygomaticus major and risorius muscles. Chimpanzees also have an open-mouthed 'play-face' expression that is equivalent to our laughter.

The fact that chimpanzees seem to smile and laugh is important, because they are our closest living relatives. It means it's likely that our common ancestor also smiled and laughed, and that the origin of these expressions goes back more than seven million years, when our ancestors were still forest-dwelling apes. It's even possible to understand how smiling would have enhanced our ancestors' survival. While in the grip of negative emotions, we tend to narrow our attention to the world around us (which could be useful when tackling short-term threats); positive emotions help us to think and behave more flexibly, something that's useful for long-term survival. Psychologists have found that people think more broadly and creatively when they're happy, and when they're smiling. But why is it useful to wear our happiness on our faces? Well, one of the hallmarks of humanity, which underlies our success as a species, is the ability to freely exchange ideas, and to produce complex culture. If smiling and laughing are also acting as 'social glue', then they'll help with this exchange of ideas. But still, we're not the only animals who smile.

ALMOST A GILL SLIT

With all this re-purposing and recycling of the contents of the branchial bars, you might think there's no obvious trace now, on the surface of your head and neck, of these embryonic structures. And indeed, the bars themselves (apart from the fifth one, which appears briefly then disappears, in a human embryo) have all become components of your jaws and your neck, and the grooves between them have smoothed over and disappeared. Except, that's not completely true. One of the grooves remains, looking more like a deep hole than a groove, admittedly: your earhole.

Your earhole, or in the anatomical jargon, your external auditory meatus, is what remains of the first branchial cleft – the groove between the first and second branchial bar. In fish, each branchial cleft grows inwards to meet its equivalent branchial pouch, growing outwards from the developing pharynx until the two meet and a gill slit is produced. Water entering the fish's mouth flows out through the gill slits, oxygenating the blood in the capillaries of the gill, which are fed by the branchial arch artery.

There's no need for any gill slits in an exclusively air-breathing animal, but there is a need for an ear – to transform sound waves in air into vibrations in fluid inside the cochlea. In the embryonic you, the first branchial cleft and first branchial pouch, on each side of your developing head, deepened towards each other, threatening to cavitate through just as they would do in a fish embryo. But although the two tunnels push close together, a membrane is left separating them, and that membrane is your eardrum.

On the inside, the branchial pouch becomes the air-filled cavity of the middle ear, and it remains connected to your pharynx by a narrow tube: the Eustachian or auditory tube. This tube is important in allowing you to equalise the pressure in your middle ear with the air pressure around your head and inside the ear canal. If you couldn't do this, then exposing yourself to low pressure, which you could do by climbing a mountain or flying in a plane, would result in the eardrum bulging outwards dangerously, due to the relatively higher pressure in the middle ear. The Eustachian tube is so narrow that it is practically closed until a relatively higher pressure in the middle ear forces a puff of air through it, making your ears 'pop'. And if your ears aren't popping on their own, you can take advantage of the fact that even adult ears and jaws are intimately associated with each other. If you open your jaws as if to yawn, you may be able to help the auditory tube to open. As the auditory tube opens into the pharynx, swallowing – which involves contractions of the pharyngeal muscles – can also help.

The other branchial clefts, below the first one that makes your ear canal, should disappear completely. But embryology is full of opportunities for things to go wrong: structures may fail to form, or they may be too large, or features which appear in the embryo and then disappear might persist. The second arch usually grows down and obliterates the second, third and fourth clefts, but if it doesn't, you're left with a narrow tube called a branchial fistula which leads into a cyst on the side of the neck. In those cases, it's almost like the remodelled branchial clefts have forgotten *not* to become gill slits.

VON BAER AND GENETICS

The branchial bars are one of the best examples of why Haeckel was wrong and von Baer was right. The human embryo doesn't look anything like an adult fish, but it does certainly does resemble a fish *embryo*, in the neck region at least. There's some reluctance to accept this theory, just in case it looks like you might believe in full-blooded recapitulation. A current undergraduate textbook of embryology kicks off its *Head and Neck* chapter with an explanation of why it will refer to the 'pharyngeal arches', rather than 'branchial arches', in a human embryo. But in doing so, it falls straight into the recapitulation trap – committing Haeckel's error of equating embryological structures in 'higher animals' with the *adult* anatomy of 'lower' forms:

> The most distinctive feature in development of the head and neck is the presence of pharyngeal arches (the old term for these structures is branchial arches because they somewhat resemble the gills [branchia] of a fish).

But it's not the gills of an *adult* fish we should be looking at, it's the branchial arches, the precursors of the gills in the fish *embryo*. Compare a human embryo with a shark embryo and it's clear that the arches in the human embryo don't just 'somewhat resemble' the shark embryo's branchial arches, they are uncannily, compellingly similar. That similarity is important because the structures really are homologous – they look the same because they *are* the same in a very real way: they are inherited from a common ancestor. The similarity goes way beyond superficial appearance too. As we've seen, it's possible to trace the tissues – the cartilages, muscles, nerves and blood vessels – that develop in each arch in the

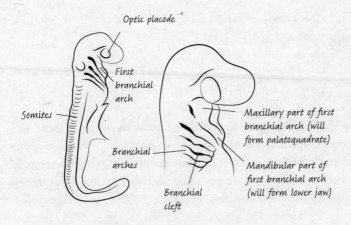

Optic placode

First branchial arch

Somites

Branchial arches

Branchial cleft

Maxillary part of first branchial arch (will form palatoquadrate)

Mandibular part of first branchial arch (will form lower jaw)

Branchial arches in a catshark

human embryo, and understand how they originated. Delving even deeper, it's possible to identify the genes which are switched on to create the segmented pattern of the branchial arches.

The cartilages inside the branchial arches come from neural crest cells which started life next to the hindbrain part of the neural tube. These neural crest cells stream down into the branchial arches, and once there, they respond to chemical signals produced by the endoderm lining the inside of the arches. The signals tell them to develop into skeletal tissue. But exactly *what* skeletal tissue they end up becoming depends on their *own* expression of patterning genes – genes which were switched on before they left the hindbrain. These genes have a very deep evolutionary history. They're called *Hox* genes, and they control the identity of segments. They're the pattern generators, not only for the branchial arches, but for all the segments of the embryo, from its neck to its tail. But if we want to understand our own *Hox* genes, we're going to have to travel back in time some 800 million years, way before any fish had evolved, to a common ancestor that we share with invertebrates. Invertebrates like the humble fruit fly.

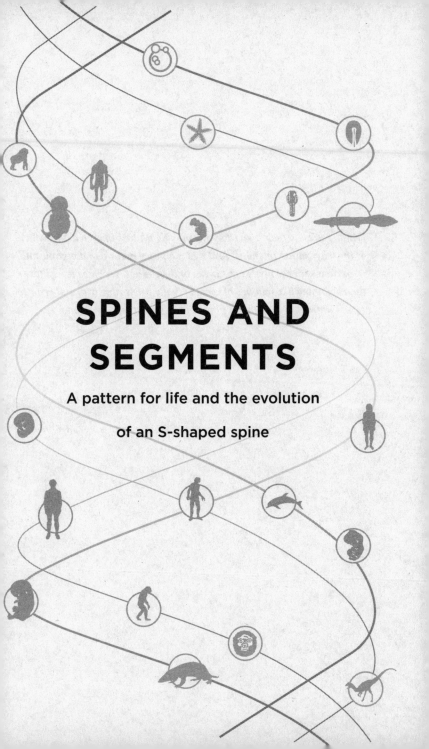

SPINES AND SEGMENTS

A pattern for life and the evolution

of an S-shaped spine

'Most species do their own evolving, making it up as they go along, which is the way Nature intended. And this is all very natural and organic and in tune with the mysterious cycles of the cosmos, which believes that there's nothing like millions of years of really frustrating trial and error to give a species moral fibre, and in some cases, backbone'

TERRY PRATCHETT

FRUIT FLIES AND THE ORIGIN OF SPINES

Each summer, the fruit bowl in my kitchen seems to become home to a small swarm of tiny flies. I assume that their eggs arrive in a batch of bananas, and then the flies settle in and start reproducing. I've never been too worried: there aren't enough of these fruit flies to call it an infestation (what is the collective term for a lesser number of fruit flies anyway? a 'mild irritation'? a 'flutter'?), and I'm actually quite fond of these diminutive insects.

During A-level biology, these flies became like pets. We learnt about genetics in practical classes, by breeding together different strains of fruit flies. Fruit flies with long wings; fruit flies with short wings. Fruit flies with white eyes and fruit flies with red eyes. Some had brown bodies, while others were yellow. They're ideal creatures for studying links between genes and observable characteristics – or 'phenotype'.

Fruit flies are famous for their contribution to genetics. It all began in the early twentieth century, in a lab in Columbia University. In 1909, the evolutionary biologist Thomas Hunt Morgan apparently grew frustrated with trying to figure out how worms regenerated and how frogs developed, and turned to studying inheritance instead, using the humble fruit fly as a model. Morgan produced mutant flies and some of the mutations turned out to be heritable, producing Mendelian inheritance patterns – including classic 3:1 ratios of dominant and recessive phenotypes – in offspring. Decades before the structure of the actual molecule which encoded the genes was decoded, Morgan's work proved that chromosomes were involved in heredity, and that genes must be arranged in a linear fashion on chromosomes. Morgan's breakthrough earned him a Nobel Prize in 1933, and the future of the fruit fly in the genetics lab was secured.

In the 1970s geneticists began to be able to sequence genes, laboriously at first, using chromatography. Then came the development of techniques which involved fluorescent labelling of the building blocks of DNA, and the whole process of DNA sequencing became automated. I've filmed in DNA sequencing labs on several occasions and the television producers have always been dismayed at how visually dull the process is: the labs are full of white boxes sitting on benches, and the magic of sequencing goes on inside them, hidden away and unaided by human hand.

When its genes were sequenced, the fruit fly revealed a truly astounding genetic secret. A particular group of mutant fruit flies had always intrigued geneticists. These were flies where body parts were swapped: sometimes segments of abdomen were swapped for thoracic segments, or the fly possessed two sets of wings, or had a leg growing out of its head where you'd normally expect an antenna to be attached. At the end of the nineteenth century, the English embryologist William Bateson had come up with a term to describe the transformation of one body part into another: homeosis. Geneticists homed in on the mutations in DNA which were producing these bizarre, mutant flies, and by doing so, they found 'homeotic genes' which appeared to control the pattern of segments in the normal fly embryo. In experiments where all these homeotic genes were deleted from a fly's genome, the result was an odd-looking larva with a row of identical segments. There was something particularly intriguing about the position of these genes in fruit fly DNA. Whereas most genes seem to be scattered about fairly randomly in the genome, so that even genes which work closely together might be on completely different chromosomes, the homeotic genes were lined up on a single chromosome – in the same order as the order of the body segments in the fly – from head to abdomen.

The homeotic genes contained a further surprise: when the genes were sequenced, each of them contained an identical, short stretch of DNA,

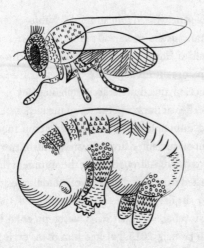

Similar Hox genes control the pattern of segments in a fruitfly and a developing human embryo

around 180 base-pairs long. This sequence was called the 'homeobox' and the genes that contained it were therefore 'homeobox genes', or *Hox* genes, for short. *Hox* genes are 'master control' genes, which work by controlling the activity of many other genes. The homeobox sequence is essential to their controlling role – it codes for a bit of a protein that can bind to DNA.

But the real surprise of *Hox* genes is that they don't just control the way that pattern is generated in a fruit fly embryo. They're in the genomes of every insect, and indeed, every arthropod. They're in worms, too. In fact, they're part of the genetic make-up of every animal with a segmented body – including vertebrates. This means that *Hox* genes are very ancient: the last common ancestor that you and a fruit fly share goes back some 800 million years. Fruit flies may be very, very distant evolutionary cousins, but the basic pattern of your body was also dictated by *Hox* genes, when you were an embryo.

Fruit flies and other insects have just one set of *Hox* genes, eight of them, lined up in a row. Our old friend, the lancelet, has a single cluster of fourteen *Hox* genes, but duplications of chunks of the genome have produced several clusters in various groups of vertebrates. Mammals have four clusters of *Hox* genes, each on a different chromosome.

The ancient nature of *Hox* genes was revealed in 1986, when a team of researchers decoded the first large vertebrate *Hox* gene cluster and found that the order of the genes mirrored the sequence of segments in the body of the animal, just as it had done in fruit flies. One of those researchers went on to demonstrate the similarity of *Hox* genes in vertebrates and invertebrates – strengthening the argument that *Hox* genes in each were inherited from a very ancient common ancestor. In the 1990s, he showed that similar genes were responsible for the patterning of limbs. That researcher was the Swiss developmental biologist Denis Duboule, and in January 2013, I visited him in his eyrie in Geneva.

We met in his office, occupying the corner of a modest tower block: the Sciences III building of Geneva University. The view through the panoramic windows took in downtown Geneva, framed by a large snow-capped bluff in one direction. The landscape of low-rise buildings (kept low by popular constraint) was interrupted by the brash excrescence of the glass-facaded television tower.

On a large table were scattered a magnificent selection of *objets de science*, including a possum skeleton (which could almost be mistaken for that of a monkey, except for the lack of a complete ring of bone around the eye socket, and the presence of extra bone sticking up from the pelvis); the skeleton of a snake, preserved in a glass bowl; phials containing tiny mouse embryos – their skeletons stained red with alizarin to highlight the bones and blue with halcyon blue to reveal cartilage; a mounted frog skeleton, and various pickled things in tall jars including a fish and a lizard of some sort.

On one wall, there was an extraordinary piece of sculpture, and I recognised its components immediately. I think almost every biologist in the world must have been sent, quite unsolicited, a volume of Harun Yahya's notorious *Atlas of Creation*. I received one, while working at Bristol University. It's a hefty tome, lavishly illustrated, and carries the creationist message that evolution is a lie: in particular, if there are animals living today that resemble fossils, this means Darwin must have been wrong. It's based on a deliberate misunderstanding of evolution by natural selection, but somehow, you can't fail to be impressed by the quality of the book – its printing and its binding – and somewhat dismayed by the thought of how much it cost to produce and distribute. My copy sat on my desk, as an oversize paperweight, for quite some weeks, then I used it as a footrest, and eventually, I consigned it to the paper recycling bin.

Denis had done something much better with a few copies of the infamous *Atlas*. On his wall hung the remains of several copies, which had been neatly sawn up into triangles and trapezoids, *screwed closed* (which, according to Denis, was a particularly French manner of dissing a book), and suspended by wires. It was a much more imaginative use for the book than my solution. It was a more thoughtful form of recycling. And wasn't there something quite evolutionary in the concept of recycling – in this case, transforming outdated ideas into art?

Denis had spent decades studying *Hox* genes, the pattern generators, or, as he put it, the architects of the body. He described how they worked, in a sequence. Rather than a gene for each segment, or group of segments, they were activated in a cumulative fashion, with more genes being switched on with each successive segment. This was how segments were determined in the fruit fly, and this was how segments were determined in your developing embryo. The most obvious place to see evidence of this segmentation in an adult human is in the spine. *Hox* genes create the differences between sections of the spine – cervical, thoracic, lumbar and sacral – as well as the differences between one vertebra and the next within those sections.

There was a glass tank in the corner of the room, and Denis opened the lid and removed a female python (a python without a name) from it. She was beautiful: sleek and long – perhaps 70cm in length. She had many more vertebrae than I did, but I was quite amazed to learn that she had the same number of Hox genes: 39.

'I find it extraordinary that she has the same number of genes as me,' I said. 'And yet so many more vertebrae. How does that work?'

'The signal for the end of the body,' he explained 'which is the position of the cloaca [the end of the gut] in a snake, is delayed, so she makes segment after segment – some 250 of them – before eventually, the *Hox* genes say "stop" and then all that is left to do is make a tail.'

There's another very obvious difference between humans and snake in terms of basic body pattern. Like most other tetrapods (amphibia, reptiles, birds, mammals), I have limbs. 'The same genes that control the segmentation pattern along the body encode for the pattern of limbs,' said Denis. 'And the signal for stopping at the ends of the fingers is the same signal that ends the body'.

We talked about the complexity of this programme of development; the fact that one gene might affect very different regions of the embryo, so that individual parts could not be freely tweaked. There were constraints upon morphology which were about making the whole structure hang together and make sense. But as well as these structural and material constraints, there were also constraints at a genetic level, because each gene didn't just do one job, in one place, at one time.

'It was one of the astounding things about decoding the human genome, realising that we didn't actually have that many genes. No more than a fly,' said Denis. 'It was also a shock to paediatricians, for instance, that different parts of the body might be controlled by the same gene. So that if there is a genetic problem with fingers or toes, you must also look at the genitalia too, because it's likely the fault will be represented there as well.'

I suspected that, while it may have been a bit of a shock, some clinicians might have expected that individual genes would turn out to affect the development of more than one body part. After all, it had been known for centuries that some inherited syndromes involve sets of defects in seemingly disparate parts of the body. But genetics now offered the prospect of tracking down the specific genes which could cause such suites of defects.

We looked at the tiny mouse fetuses in their phials, made translucent and stained red and blue to light up the skeleton. This was how Denis and his colleagues could pinpoint the genes which controlled particular aspects of development: 'The experimental work is painstaking,' said Denis. 'It takes two years to make the genetic change you want, and to let a mouse embryo develop so that you can see the effect of changing a gene. Not many students are prepared to spend that long waiting for results.'

But the lengthy work had paid off; the resultant embryos showed by their defects what the normal gene's role would have been in the developing embryo. And in a system as highly conserved as the *Hox* genes, the development in humans was very likely to be similar to that in mice. Biologists like Denis were getting closer to understanding the programme which created the body pattern of vertebrates, but also gaining a better understanding of how the programme could go wrong.

Thomas Hunt Morgan began his scientific career investigating development and regeneration, looking at how patterning could be generated in embryos and regenerating limbs, before switching his focus to inheritance. He'd have been pleased that his model organism of choice – the fruit fly – has had such a huge impact on the field of genetics, and not only that, but the fruit fly would help to solve the riddle which had so frustrated him, illuminating the links between genes and embryology.

EMBRYOLOGICAL DEVELOPMENT OF VERTEBRAE

So this is where we return to the tiny embryonic you, just four weeks after conception, when you were less than half a centimetre long. You have those chordate characteristics: a notochord, a hollow neural tube and a tail. You also have branchial arches, like all vertebrate embryos. On the sides of your tiny head, the thick, round placodes which will become the lens in each of your eyes are sinking inwards. And all along your embryonic back there is a double row of bumps, like two strings of beads.

These bumps have appeared in response to internal signals, produced by the cells inside the bumps themselves. But then other tissues start to talk to the bumps, in the language of chemical signals or morphogens, of course. And as those morphogens diffuse into the bump from surrounding tissue, the cells inside the bump begin to differentiate. In pre-vertebrates like *Haikouella* and lancelets, the notochord sticks around as an important strengthening rod down the axis of the body, but in vertebrates like you and me, it's replaced by a series of bones, assembled with flexible joints between them, to form a spine. In a way, the notochord has a role to play in its own demise, because it is one of the tissues producing signals which tell cells in the bumps (which we should really call somites) to form parts of the skeleton – including vertebrae. Other signals, seeping out from the neural tube, tell cells in the upper, outer part of the somite to become precursors of the dermis (the lower layer of the skin, under the epidermis) and muscles.

Everything that derives from one particular bump, or somite, is supplied by a single spinal nerve. Wherever those tissues end up, they 'remember' their segmental origin in their nerve supply. Muscles might end up migrating a long way away from their original segments, but they *still* maintain their original nerve supply. For instance, some muscle cells from the fifth and sixth cervical somites end up migrating into the developing upper limb bud, to form a muscle called biceps brachii (biceps to its friends). So your biceps muscle, in each arm, is supplied by a nerve which you can trace all the way back to the neck, to the fifth and sixth cervical spinal nerves, emerging from the spine.

In each somite, the cells of the inner, skeletal component migrate towards the middle of the embryo, and meet cells from the somite on the other side, above the neural tube and around the notochord. These cells eventually condense into a cartilage model of a vertebra. In fact, each single vertebra is formed by the lower half of one somite fusing with the upper half of the next somite below it. This is important, as it means that the spinal nerves, exiting at a level which is right in the middle of a somite, will emerge between one vertebra and the next. It also means that the muscle blocks developing in the outer portion of each somite will bridge between one vertebra and the next – meaning that those muscles will be able to move the spine.

There are some things you can do to make sure that the spine gets off to a good start, if you're an expectant mother. The cells from the somites that gather around above the neural tube will eventually form the neural arches of the vertebrae, which enclose and protect the spinal cord. If the cells fail to meet each other in the middle, the neural arch will fail to form properly, and the spinal cord will be left unprotected. This condition is called spina bifida (literally: 'divided spine'). In the most mild cases, a vertebra is split at the back, but the spinal cord and the skin over it is intact, and there are no problems with spinal nerves. In more severe cases, nerve tissue from the spinal cord is exposed on the surface in the middle of the baby's back. This lays the spinal cord open to infection, but it also means that there are likely to be serious problems with the nervous system. Even if the defect can be surgically closed, the baby will probably have paralysed legs.

One of the risk factors for spina bifida is a low level of folic acid in the mother. Folic acid is an essential vitamin which has an everyday role to play in DNA synthesis, as well as being important during neural tube formation in the embryo. We can't make folic acid in our bodies, we have to obtain it through our diet. It's found in abundance in green, leafy vegetables (which is where its name comes from: 'folia' means 'leaf' in Latin) as well as in peas and beans, egg yolk and sunflower seeds. In countries where women are encouraged to take folic acid supplements, the rate of neural tube defects has dropped significantly. Some cereals and bread are also fortified with folic acid. Studies estimate that, since the introduction of folic acid-fortified cereals in the US in 1998, the rate of neural tube defects like spina bifida has been halved. In fact, it's estimated that up to 70 per cent of these birth defects could be prevented by mothers taking folic acid supplements early in pregnancy. This is why it's important for all pregnant mums – and indeed, any women who are hoping to conceive – to make sure they have enough folic acid, and the best way of doing that is to take folic acid supplements, alongside a healthy, balanced diet.

Although most vertebrae conform to a very general stereotyped shape, there are differences between them. The identity of different vertebrae in the spine is determined extremely early in embryological development, with *Hox* genes being switched on in cells along the axis of the embryo, even before somites start to form. The pattern of *Hox* genes switched on in each segment tells the bone developing out of the sclerotome what type of vertebra it should become: cervical, thoracic with ribs attached, lumbar, sacral with pelvis attached, or caudal – for the coccyx.

ANATOMY OF THE SPINE AND SPINAL CORD

There are some features of vertebrae which are shared by most of them, others which are characteristic of particular regions (like cervical or lumbar), and then others still which are quite individual. The general features of a typical vertebra reflect its two main jobs: in order to support the body, it must bear weight, and it must also protect the spinal cord. The weight-bearing part of a vertebra is its body, and behind the body there is an arch which encloses the spinal cord. There are various bony prongs (called processes) extending from the arch which are levers for muscles to pull on. At the back of the spine, the arches overlap like flat, armour plates, protecting the cord. Sticking something into the space between the arches is difficult, but that's exactly what an anaesthetist needs to do to deliver an epidural anaesthetic.

With the patient lying on their side, curled into a fetal position, or sitting on a bed and curling forwards, the gaps between the bony arches are opened as wide as they can go. It's still not that wide, but enough to let a needle in. The linings of the spinal cord are the same as the linings

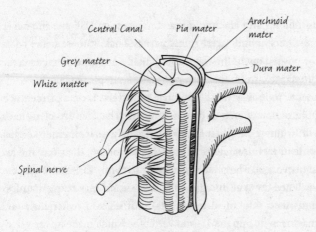

Central Canal   Pia mater   Arachnoid mater

Grey matter

White matter   Dura mater

Spinal nerve

The layers of meninges around the spinal cord

around the brain: three layers of meninges, starting with the pia mater, stuck closely to the spinal cord itself; the fine, cobwebby arachnoid mater; finally, the tough, thick covering of the dura mater. The space between the dura mater and the bony neural arch is full of fat and lots of veins. This is the epidural space, where the anaesthetist's needle is headed. (Epidural means 'around the dura'.) When anaesthetic fluid is injected into that space, it affects the spinal nerves emerging from the cord. The spinal cord, like the brain, lies bathed in cerebrospinal fluid (CSF) which occupies the space beneath the arachnoid lining. In a spinal anaesthetic, the needle is pushed in further, through the dura and arachnoid mater, into this space, so the spinal cord itself is anaesthetised. This is also the space that a needle must be pushed into in order to withdraw a sample of CSF in a lumbar puncture, but this is always done below the end of the spinal cord. Although babies are born with spinal cords which are almost as long as the spines they sit in, the vertebral column grows in length more quickly, until the spinal cord ends up tapering to a stop, high up in the lumbar spine, level with the intervertebral disc between the first and second lumbar vertebrae. Below the spinal cord, there is a long 'bag' of dura and arachnoid mater, containing CSF and the nerve roots of the lumbar and sacral spinal nerves. Luckily, these nerve roots tend to be pushed out of the way without being damaged when a needle is introduced into the lumbar cistern to take a sample of CSF.

It's possible to look at most vertebrae as 'variations on a theme', each with a body, a neural arch and some processes, but nevertheless, each one is a little different. When I'm looking at an archaeological skeleton in the lab, the first thing I do is remove the bones from their box and lay them out in order on a long table. I usually start with the bones along the axis of the body: the skull and the spine. The skull is easy enough (unless it's been crushed into small fragments), but it's also easier than you might imagine to line up the 33-odd vertebrae which make up the vertebral column. Firstly, I'm helped by the fact that the bodies of vertebrae get progressively larger as you go down the spine – each vertebra is bearing more weight than the one above. But there are also important differences in vertebrae from the cervical (neck), thoracic, lumbar, sacral and coccygeal regions.

A typical cervical vertebra has a small body, a large neural arch, a forked spine and, in each of the transverse processes (prongs sticking out to the side), a hole. This is the hole that the vertebral artery passes through, running up to the neck before finally entering the skull and joining up with the internal carotid arteries to supply the brain with oxygenated blood.

There are two cervical vertebrae that are quite different from the rest, though, and these are the top two vertebrae in the spine. The main reason that they are odd comes down to a few joints which are very important to the way you move your head to communicate. The uppermost vertebra is called 'atlas' (after the ancient Greek Titan who held up the sky) and this vertebra supports the skull. There are two large, curved facets on the atlas which form joints with the lower part of the occipital bone, either side of the foramen magnum where the brainstem exits the skull. These joints allow the skull to rock forwards and backwards on the spine: they let you nod. The atlas is a very odd vertebra indeed; it is unique in having no body. Where the body of the atlas should be, a bony peg sticks up from the second neck vertebra, which also has a specific name: the axis. The peg of the axis makes a joint with the arch at the front of the atlas, so that the atlas (together with the skull above it) can rotate from side to side: this is the joint which allows you to shake your head.

And then, all the way down the cervical spine, there is movement between each vertebra and the next. You can flex your neck forwards and bring your chin down to your breastbone, extend it backwards until you are gazing at the sky (or the ceiling), and flex it to each side as well. You can add all those movements together to roll your head around on your spine.

Thoracic vertebrae are characterised by attachments for ribs. There are small facets on the sides of the vertebral bodies, and on the ends of the transverse processes. Lumbar vertebrae are chunky, with large bodies compared with the size of the neural arch, and their 'spines' are square plates of bone. Movement is limited in the thoracic spine, which is braced by the ribcage, but freer again in the lumbar spine. Then there's next to no movement right at the end of your spine, in the sacrum and coccyx.

Your anatomy reflects your evolutionary heritage and your kinship with other animals, and your spine is no exception. There will be broad similarities when you compare yourself with distantly related creatures. As a vertebrate, you have a spine, and that's a basic similarity you share with fish. The job of the spine remains the same across all vertebrates: to provide a strong, flexible support along the length of the body, and to protect the precious spinal cord. Comparing yourself with other vertebrate land animals (in the jargon, tetrapods) you find that you share a spine with different sections: cervical, thoracic, lumbar, sacral and caudal – although, as with Denis Duboule's snake, the numbers of vertebrae may vary considerably. If we focus even more narrowly, on our closer mammal cousins, the numbers of vertebrae turn out to be relatively conservative for the whole group. The most variation is in the tail. Some animals (like us) have just a handful of caudal vertebrae.

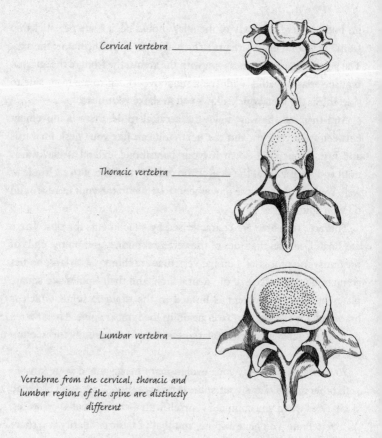

Cervical vertebra _____

Thoracic vertebra _____

Lumbar vertebra _____

Vertebrae from the cervical, thoracic and
lumbar regions of the spine are distinctly
different

The fast-swimming Dall's porpoise (*Phocoenoides dalli*) probably wins
the accolade for having the most caudal vertebrae of any mammal, with
up to 49 in its tail. The least variable region is the neck: all mammals,
even porpoises and giraffes, have seven cervical vertebrae.

Despite a small amount of variation where the boundaries between
regions shift – a neck vertebra gains a rib, a lumbar vertebra becomes
attached to the sacrum – we humans usually have seven cervical vertebrae, twelve thoracic vertebrae (with twelve pairs of ribs accompanying
them), five lumbar vertebrae, five sacral vertebrae and between three

and five small caudal vertebrae, which may be fused together, forming the tailbone or coccyx. These last few vertebrae in the human spine are named from the Greek for 'cuckoo', because it's meant to be shaped a bit like a cuckoo's beak. They are sometimes said to be vestigial, merely a remnant of a tail that our ancestors once had, but which we could do without. But trust me, you wouldn't want to be without a coccyx. Although it's not much of a tail, it forms a very important anchor point for ligaments and muscles, including your levator ani, better known as the pelvic floor. Without a coccyx, there'd be little hope of keeping the contents of your pelvis from dropping out onto the floor.

SLIPPED DISCS AND FAULTY FACETS

Your vertebrae above the sacrum are joined together by a peculiar sort of joint which resists compression and absorbs shock while at the same time allowing movement: the intervertebral disc.

During the embryological development of the spinal column, the cells in the 'gap' between one vertebra and the next form fibrocartilage, creating a flexible joint – the intervertebral disc. The notochord is obliterated where the body of a vertebra forms, but notochord cells persist for some time between the vertebrae in the pulpy centres of the intervertebral discs. Here, they may play a couple of important roles: they're undifferentiated stem cells, so they could divide to replace cells in the intervertebral disc. They may also regulate the function of other cartilage cells in the disc. But it seems that the notochordal cells may be gradually lost from the intervertebral disc as you get older, and this could be one of the reasons why discs are likely to degenerate with advancing age.

As they degenerate, your discs are quite likely to cause you some trouble. The intervertebral discs in a young spine do their job very well: they resist compression and act like cushions between the vertebrae. The gel-like centre, which is the consistency of toothpaste in a young disc, helps to spread load across the ends of vertebrae, and it's kept in place by a tough, fibrous ring. But sometimes the gel centre can herniate out through cracks in the fibrous ring, producing what is commonly known as a 'slipped disc'. Because the spinal nerves exit the spine between one vertebra and the next, this means they are level with the intervertebral discs. Also, when the disc does herniate, it tends to do so where the pressure is high, towards the back – just where the spinal nerves lie. This can cause lower back pain, but sometimes the pain is felt further away, along the course of a nerve that is being compressed by a slipped disc. This type of problem tends to affect the lower back, probably because it's bearing more weight than higher levels. The spinal nerves exiting between the lower lumbar vertebrae run down into the pelvis and join up with sacral nerves to make a huge nerve – up to 2cm wide – which runs down the back of the thigh: the sciatic nerve. So this is how a slipped disc in the spine can cause shooting pains down the back of the leg, known as sciatica. The gel centre of the intervertebral disc also tends to shrink and dry out with age, meaning that loads are not spread evenly across the vertebral endplates, which can cause micro-fractures. Sometimes, rather than herniating out through the fibrous ring around the edges of the disc, the gel can herniate up or down, pushing into the vertebra itself. This pocket of interverte-

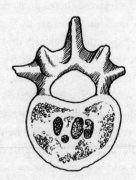

A Schmorl's node in a vertebra from an archaeological site

bral disc breaking through the endplate of a vertebra is called a 'Schmorl's node'. These nodes are visible on x-rays, as dark blobs near the top or bottom of a vertebral body. They are also easy to spot in dry bones: with the disc long gone, the hole that it punched into the vertebral body is still visible.

Human spines are 'good enough' as far as evolution is concerned, but plagued with problems, especially in later life. If you've ever suffered from back pain, you're certainly not alone. Lower back pain is the most common complaint seen by GPs, and it's estimated that about two-thirds of people in the general population suffer from spinal pain.

Each of your intervertebral discs allows a little movement, a bit of rocking and rotation, between one vertebra and the next. But there are also more 'conventional' joints between vertebrae which, like your hip, knee or finger joints, are lined with smooth, hyaline cartilage and lubricated with slippery, synovial fluid.

These facet joints allow movement in certain directions, but what's more important is the way that they limit movement in the spine. In the chest, where the vertebrae are largely held in place by the ribs, the facet

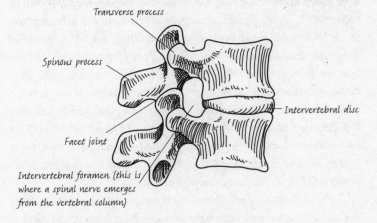

Transverse process

Spinous process

Intervertebral disc

Faeet joint

Intervertebral foramen (this is where a spinal nerve emerges from the vertebral column)

The joints between the vertebra include the intervertebral disc and the facet joints

joints allow a little forwards flexion and backwards extension, but almost no rotation. In the neck, rotation is much freer. In the lumbar spine, the facet joints allow flexion and extension and a bit of rotation. In fact, compared with our closest ape cousins, our lumbar spines are very flexible indeed.

Moving parts tend to be the most troublesome elements of any engineering structure, and the joints in your spine are no exception. We've seen how the gel nucleus of the disc can be squeezed out and cause problems, but the facet joints are another source of back pain. Even when you think you are sitting or lying very still, you're probably still shifting your weight around a little, moving your spine, so your facet joints are rarely at rest. Over a lifetime, they bear the burden of repetitive strain and minor injuries. They're prone to degeneration, even more so as the discs themselves degenerate with increasing age. Weight tends to get shifted backwards onto the facet joints, which are not really designed for that much load-bearing. Eventually, the hyaline cartilage surfaces are worn away, and spurs of bone grow around the edges of the joints – classic signs of osteoarthritis.

The facet joints are supplied with sensory nerves which branch off from the spinal nerves leaving the spinal cord through the gaps between the vertebrae, so the joints themselves can be a source of pain. But how much back pain could be attributable to these joints? A group of American anaesthetists set about finding out by injecting anaesthetic into the facet joints of 500 patients suffering from back pain. They discovered that the anaesthetic numbed the pain in around half the patients who were suffering from chronic neck pain, slightly less than half of those with pain in the thoracic spine, and about a third of those with chronic pain in the lumbar spine. The facet joints also turn out to be the source of pain in a significant number of people suffering chronic problems after whiplash injuries to the neck.

The facet joints illustrate something about the nature of pain itself: it's not always easy to work out where pain is coming from. If you damage your outer surface, you can usually pinpoint the source of the pain with precision. Remember the last time you were bitten by a mosquito? It's a minute injury, but you know exactly where the bite is happening – usually in time to eliminate the offending insect with a slap. The outside of your body is mapped onto the sensory cortex of your brain, each bit of incoming information is effectively plotted onto the map.

It seems to be much harder to locate the source of pain emanating from internal organs. Your insides are supplied with a different set of sensory neurons; ones which usually provide information to the brain at a subconscious level, allowing you to control various aspects of your physiology without having to spare a conscious thought for them. Indeed, if this wasn't the case, you wouldn't have time to think about anything other than your internal workings. You don't need to consciously increase your heart rate when you run, or intentionally empty your gallbladder into your duodenum when you have a meal. That's fortunate, because if you were busy thinking about all this house-keeping, you'd hardly be able to have a conversation, solve a problem or read a book. But when the sensory signals coming from your insides reach a particular intensity, they do start to emerge into consciousness. You become aware of discomfort or pain, but it's very difficult to know exactly where it's coming from. Sometimes the ache is diffuse and nebulous. At other times, it's as though the brain gets its wires crossed, and the pain is interpreted as coming from a structure far away from the actual source. This is a phenomenon known as 'referred pain'. It seems to be particularly difficult for the brain to sort out signals from different sensory nerves entering the spinal cord at the same level, and this all goes back to the original segmentation of your body as an embryo.

Each of those somites (those bead-like bumps along the back of the developing embryo) contains groups of cells within it which are destined to be skeleton, muscle and dermis. No matter where those tissues end up, they will 'remember' their origin in the somite through the spinal nerves that supply them. For example, the fourth spinal nerve (L4) will supply parts of vertebrae, along with segments of particular muscles in the leg, and skin over the lower back and leg that all came from the fourth lumbar somite in the embryo. The pain from a facet joint supplied by L4 may not be felt in the back, precisely where the problem lies, but might be felt across the lower back and buttock or back of the thigh, and sometimes even the lower leg and the foot – areas which are also supplied by L4.

Injection of local anaesthetic into a lumbar facet joint may also be a useful way of both diagnosing and treating pain in these joints. If you have lower back pain, it turns out that one of the best things you can do for yourself is to take up yoga.

Perhaps if we had much more stable lumbar spines, like chimpanzees or gorillas, we'd be largely spared the burden of lower back pain. But flexible lumbar spines are part of what makes us human: they're very important to the way we move around.

LONG LUMBAR SPINES

Chimpanzees and gorillas usually have the standard seven cervical vertebrae, thirteen thoracic vertebrae, four lumbar and six sacral vertebrae. So, compared with the African apes, we humans seem to have a relatively long lumbar spine. But it's nowhere near as long or as flexible as that of our more distant primate relatives, the Old World monkeys of Africa and Asia, for whom the normal number of lumbar vertebrae is seven.

If we delve into our evolutionary past, we find ancestors who had long, mobile lumbar spines, just like those of living Old World monkeys. Fossils of *Pliopithecus* show that fossil monkeys, which lived around fifteen million years ago, had flexible lumbar spines comprising six or seven vertebrae, along with a short sacrum and a long tail. So this raises two questions: when did our spines get so much shorter? And then, why aren't they as short as chimp or gorilla spines?

It turns out that shorter lumbar spines were part of the anatomical changes which happened as apes evolved from monkeys, along with an almost complete loss of tail vertebrae. The loss of a tail is a bit of a conundrum, as long tails are so very useful to animals living in trees – to help with balance. When large monkeys like baboons walk along horizontal branches, on all fours, they whip their tails to the side to recover their balance. The earliest apes would have moved along branches in a similar way, on the palms of their hands and the soles of their feet. But early fossil apes like *Proconsul* have already lost their tails. While they seem to have made up for it with more mobile limbs, well-stabilised elbows, and hands and feet that are even better at grasping than those of their monkey predecessors, the reasons for the initial loss of a tail are still somewhat mysterious. I'm proud of my ape heritage, but I do sometimes feel that it's sad that evolution has denied us this bit of anatomy that most of our primate cousins, from ring-tailed lemurs to spider monkeys to baboons, still enjoy.

For a large-bodied animal, with or without a tail, there's a point at which walking on all fours along the top of a branch becomes too risky. It's much safer to hang your weight under the branch, or to stand up straight and walk along a branch on your legs, using your hands to support you. This creates a fundamental difference between monkeys and apes: monkeys tend to move on all fours in the trees, with their trunks horizontal; apes tend to clamber, climb, hang and swing with their bodies upright. And if you're an ape moving in this way, having a flexible lumbar spine suddenly becomes a liability. A mobile spine would be prone to injury, while a stiff back would help to support your whole trunk as you bridge between trees, standing in one and reaching out to the branches of another. Now it becomes clear why modern chimpanzees and gorillas have short lumbar spines. And this segment of their vertebral column is not just shorter, it's very stable. The upper part of the pelvic bones rise up high on each side, 'trapping' the last lumbar vertebra so that it can barely move.

Although you *are* an ape, your lower spine and pelvis looks entirely different. You've got a longer lumbar spine that's no longer entrapped by the pelvis on each side. There's also an important difference in the shape of the bodies of your lumbar vertebrae. Looking at your lumbar vertebrae from the side, it's clear to see that each one is wedge-shaped – taller

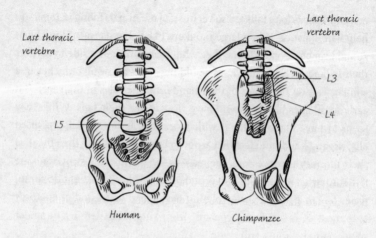

Last thoracic vertebra

Last thoracic vertebra

L3

L4

L5

Human

Chimpanzee

A human lumbar spine contains five vertebrae compared with four in the chimpanzee, and the lower lumbar vertebrae in a chimpanzee are trapped between the high ilia of the pelvic bones

at the front than at the back. This means that these vertebrae naturally stack up to form a backwards curve, called a lumbar lordosis. This characteristic inwards curve in the small of your back is something that other apes don't possess. This wedging of lumbar vertebrae was already present in some very ancient members of our hominin family tree. A well-preserved fossil spine of *Australopithecus africanus* (known rather lyrically as STS 14), a species which existed between two and three million years ago, shows that the lumbar vertebrae were wedged. Proof that these early hominins had a lumbar lordosis, like us. These early hominins also have long lumbar spines, with five or even six vertebrae.

Having an extra curve in your lower spine might not seem exactly mind-blowing – you may be thinking, so what? But the backwards bend in your lumbar spine means that you can bring your upper body upright, balancing your chest over your pelvis, and your pelvis above a pair of straight legs. Chimpanzees and gorillas seem to find it difficult to bend their spines back to move their body weight over the pelvis, so when they stand up and walk on two legs, they tend to do it with bent hips and bent knees. They can walk on two legs, certainly, but not as efficiently as us. Try walking around with bent knees and hips for a bit, and you'll feel your thigh muscles having to work hard. If you're a habitually bipedal ape (that's you, me, all humans and our hominin ancestors), having a backward bend in the spine is an essential part of making standing and walking on two legs efficient. So the lumbar lordosis in australopithecine spines shows that these ancestors were already committed to regularly walking upright on two legs. There's another side to this coin: it's been argued that if a short, stiff spine reduces the risk of injury and helps with moving between trees, then the long, flexible lumbar spine of *Australopithecus africanus* suggests that these hominins had virtually abandoned life in the trees.

You weren't born with a lumbar lordosis. To begin with, the spines of newborn babies naturally curl into a single, forwards curve. When a

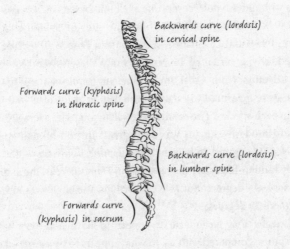

Backwards curve (lordosis) in cervical spine

Forwards curve (kyphosis) in thoracic spine

Backwards curve (lordosis) in lumbar spine

Forwards curve (kyphosis) in sacrum

The curvaceous human spine

baby is about a year old, the backwards curve in the lumbar spine starts to develop. Another backwards curve develops in the neck, to balance the weight of the head above the trunk. Eventually, the human spine ends up with a beautiful 'double-S'-shaped curve, with a backwards curve in the neck, a forwards curve in the thorax, another backwards curve in the lumbar spine, finishing off with the sacrum curving forwards under it.

A STRAIGHT-BACKED COUSIN

The story of the lumbar spine takes on an odd twist (metaphorically speaking) when we look at the vertebral columns of our closest cousins within the hominin family tree: Neanderthals.

These stocky humans lived in Europe and Asia, between about 300,000 and 30,000 years ago. There's a huge amount of fossil evidence of Neanderthals – the remains of more than 275 individuals have been discovered – so although no one skeleton is complete, it's possible to build a whole 'composite skeleton' based on remains from various sites. Neanderthal fossils from the sites of Shanidar in Iraq, Kebara in Israel and La Chapelle-aux-Saints in France all include well-preserved lumbar vertebrae. In a recent study, these vertebrae, together with a selection from other, more ancient hominins, and modern humans, were measured to establish the degree of lordosis in individuals from each species. The results were surprising. The average lumbar lordosis in modern humans was 51 degrees. In non-human apes the average was 22 degrees. In ancient *Australopithecus africanus*, the angle was fairly human-like, at 41 degrees, and in *Homo erectus*, the angle was a respectable 45 degrees. But in Neanderthals, the angle was low, averaging a mere 29 degrees.

This means that Neanderthals appear to have had very straight lumbar spines, compared with us. In what appears to be a reversal of the trend towards more curvy spines in human evolution, our close cousins

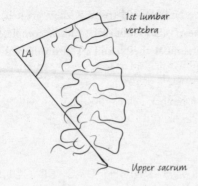

1st lumbar
vertebra

LA

Upper sacrum

The lordotic angle (LA) of the lumbar spine (redrawn from Been et al. 2012)

had developed straighter backs than their ancestors. Not only does a curved lumbar spine make for an efficient upright posture, the addition of a springy curve also helps to act as a shock absorber. Without a lumbar lordosis, it's likely that Neanderthals would have walked somewhat more slowly, with a shorter stride and a trunk bent slightly forwards. This seems to be a regressive step, but there could be a very good reason for having a straight spine – and that is stability.

A straight, strong lumbar spine might have reduced the stress on the vertebral column, perhaps allowing Neanderthals to do more rigorous upper body activity and carry heavier loads than their slighter cousins, our modern human ancestors. Who knows, if we'd descended from Neanderthals (some of us, including me, have a small percentage of Neanderthal DNA in our genomes, but apparently not enough or not in the right place to change our spines), maybe we wouldn't be so plagued with lower back pain today.

Although Neanderthal skeletons are very similar to ours in many respects, it seems that there were some important differences, which probably reflect differences in lifestyle between them and our own ancestors. Along with straighter lumbar spines, another striking feature of the Neanderthal trunk is their great barrel-shaped chests. But your chest is also unusual – it's a completely different shape from most mammal chests, and it's also different from the chests of other apes.

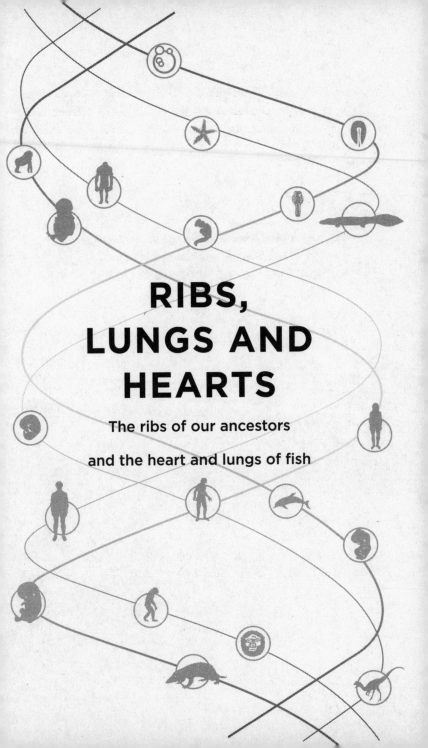

RIBS, LUNGS AND HEARTS

The ribs of our ancestors

and the heart and lungs of fish

'Knowst thou how blood, which to the hart doth flow,
Doth from one ventricle, to th'other go?'
JOHN DONNE, *OF THE PROGRESSE OF THE SOULE*

THE RIBCAGE AND DIAPHRAGM

When I look at an archaeological skeleton, the first thing I do is lay out the bones in an anatomical arrangement, as though the individual was lying on their back, arms by their sides, palm uppermost. Then I make an inventory of the bones, before moving on to look at each bone more carefully, taking note of features which might help me to determine the age and sex of the individual, as well as any telltale signs of disease.

Ribs can be a real pain: they're often broken into short fragments, but after some patient work on this jigsaw, it's possible to put them in order. A human chest is shaped like a barrel which has been squashed front to back, and the shape of individual ribs reflects their position. The facets around the head and neck of each rib help to reveal whether it comes from the right or left side of the chest. The uppermost and lowermost ribs are also quite distinctive. The first rib is a small, C-shaped bone, flattened from top to bottom, whereas all the other ribs are flattened side to side.

Like most of the other bones in your body, ribs develop from cartilage models which themselves develop out of embryonic connective tissue or mesenchyme. The cells which form ribs come from the same source as those which make the thoracic vertebrae: from somites (that row of bumps along the back of the embryo) in the chest region of the embryo. And just as there are twelve thoracic vertebrae, there are twelve pairs of ribs (and exactly the same number in both sexes, despite what some creationists believe). The breastbone or sternum develops out of mesenchyme which starts off on the outer edges of the embryo while it's still a disc. When the embryo rolls up into a cylinder, in the fourth week of gestation, these edges are brought together in the midline, at the front. Just like the ribs, a cartilage model of the sternum develops first, then later turns into bone.

The upper seven pairs of ribs in your ribcage reach from the vertebrae at the back, right around to meet the sternum at the front. Actually, that's not quite true, because the bony part of the rib ends short of the sternum, with a bar of cartilage completing the arc: the costal cartilage. But in fact, as the ribs all started off as cartilage which later turned to bone, the costal cartilage is really just the unossified front end of the rib. They give your ribcage flexibility, something you may have been glad of if you've ever got stuck at the front at a lively rock concert. It's also what makes chest compressions possible in cardiopulmonary resuscitation (CPR): the breastbone can be pushed down some 5cm, to squeeze blood through the heart and keep the circulation moving until someone with a defibrillator arrives on the scene.

Your ribs are joined together by intercostal muscles, and each set of these muscles is supplied by branches from one pair of spinal nerves. The segmental innervation of these muscles in your chest is another reminder that your body is essentially segmented – that's the way it develops in the embryo, just as in the larvae of fruit flies.

Many anatomy textbooks still describe the main function of the intercostal muscles as to move the ribs to expand or contract the ribcage when you breathe in and out. While the intercostal muscles can certainly produce changes in the shape and volume of the thorax, that's not their primary function. And, of course, you've got a much more important muscle when it comes to respiration: your diaphragm. This dome-shaped muscle attaches around the lower margins of your ribcage, to the spine at the back, and to the bottom of the sternum at the front.

If you're sitting quietly right now, your diaphragm is probably moving up and down about a centimetre and a half, perhaps twelve times a minute, keeping an ebb and flow of air moving in and out of your lungs like a tide. Your diaphragm domes up high into the chest cavity when it's relaxed, but if you now breathe in deeply, it will contract and flatten until it lies about 10cm lower. Inside your chest, the volume of your lungs has increased, meaning that the pressure inside them drops, sucking in air from the outside, through your nostrils, your nasal cavity, your pharynx, larynx, trachea and bronchi. The main role of the intercostal muscles, meanwhile, is not to move the ribs, but to resist the tendency of the spaces between the ribs to get sucked inwards by the negative pressure inside your thorax as you breathe in. The intercostal muscles stiffen the chest wall, making sure that the thoracic volume gained by flattening the diaphragm isn't lost by the intercostal spaces getting pulled inwards.

The contraction of the diaphragm can increase the volume of your thorax (and your lungs inside) by up to 3 litres, but only when the intercostal muscles are working properly. If the intercostal muscles are paralysed, they get sucked inwards with each breath, and about half of the normal increase in lung volume is lost. This can happen with injuries to the spinal cord low down in the neck. Damage at this level will affect all the spinal nerves below the injury, including the nerves supplying the intercostal muscles. But the diaphragm can still work – and that's because of yet another quirk of anatomy which we can only understand by looking back at embryonic development. On each side of your neck, there's a phrenic nerve running down, passing under the clavicle, under your first rib, and into your chest. It runs down, next to your heart, all the way to your diaphragm, which forms the lower boundary of your thorax. If we delve back into your embryonic past, going back to the fifth week of your existence, we find that your embryonic heart has already formed, and it's even pumping away, but it lies right up in the neck region. Below the heart, but still up in the neck at this point, a broad wedge of tissue is growing from the front towards the back of the embryo. This transverse septum is the precursor of the diaphragm, and the branches of cervical spinal nerves which grow out to it at this early stage will stay faithful to it. Over the next month, the developing diaphragm will move downwards, as the embryo grows in length. Eventually, the diaphragm will end up at the bottom of the thorax, but it will keep its nerve supply, coming from much higher up, in the neck. That's why, as medical students learn, 'C3, 4, 5, keep the diaphragm alive'. The phrenic nerves originate up in the neck, formed from the third, fourth and fifth cervical spinal nerves.

THORACIC SHAPES

Ribcages in different mammals come in a great variety of shapes and sizes but they must all fulfil the same functions as far as the organs on the inside are concerned. The ribs protect the heart and lungs, and provide attachment for the diaphragm and intercostal muscles, allowing breathing to happen. But the shape of the chest isn't just about what's inside, but what's connected to it on the outside. Crucially, that includes the shoulder blade, which connects to the forelimb in other animals, the upper limb (arm) in us. Animals which walk and run on all four legs tend to have a ribcage which is flattened from side to side. Think of a domestic cat or dog: their chests are narrow and deep. Monkeys, which are also quadrupeds, have narrow thoraces too. In all these animals, the shoulder blade is mounted on the side of the narrow chest, and it rotates as part of the movement that brings the forelimb forward and back in each stride. But apes are different; the lower part of the ribcage is wider than it is deep, and it's flattened front-to-back. Humans are no exception, we all have front-to-back flattened chests – you can check this in yourself. Vets will routinely take chest x-rays of dogs and cats from the side, whereas the standard x-rays which doctors look at show a view of the human chest from front to back. (Actually, these standard x-rays are taken the opposite way round, so they're known as postero-anterior or PA views.)

This chest shape is related to the special way in which apes move around. Apes evolved from smaller-bodied monkeys, but they're certainly not just scaled-up monkeys, and they don't move around like most monkeys, either. Walking on all fours along the top of a branch gets tricky when you're big. It's much easier and safer to climb and clamber, reaching your hands out to branches above your head for support. You can climb out along quite slender branches like this, and get right out to the edges of trees, where fruits are most abundant – and that's another characteristic of apes: most of them are frugivores. Some living apes, including gibbons and orangutans, also suspend themselves from their arms, again placing their trunks in an upright position.

There are some monkeys which have distinctively 'ape-like' chests, flattened from front to back: the spider monkeys (*Ateles*) of Central and South America. The shape of these monkeys' chests is an example of convergent evolution – they move around in a similar way to gibbons, swinging through trees and hanging from branches, with an upright trunk. Their chests are adapted to this in the same way as apes'. And of course there are some apes which no longer move around in trees (much) but still have front-to-back flattened chests: us. This chest shape is inherited from ancestors who spent much more time climbing in trees, but it's also very well suited to an ape which walks around on the ground, with an upright chest. But there are also some differences between our chests and those of other apes.

Comparing the shape of a human chest with that of an orangutan, there are obvious differences. The human chest is barrel-shaped, with the bottom of the ribcage curving back in. But the orangutan's chest is funnel-shaped, with the lower ribs flaring out. The orangutan's chest is narrow at the top, and the shoulder blades lie close together and high up; orangutans and other non-human apes look as though they're perpetually shrugging their shoulders.

In contrast, the top of a human chest is quite wide, and the shoulder blades are lower and further apart. This might not place the shoulders into such a good position for climbing, but it means you can swing your arms to help with counterbalance as you stride along. The next time you're out and about, try walking with a nice, long, relaxed stride and notice how your chest, shoulders and arms move. Your shoulders twist in a counter-rotation, against the twist you're introducing by swinging a

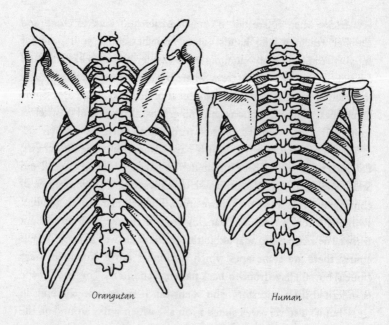

Orangutan

Human

A funnel-shaped orangutan ribcage and a barrel-shaped human ribcage

leg forwards. Your arms swing to help this twist. If you try keeping them level, walking suddenly feels much less comfortable, and in fact it's also less energetically efficient. With a chest that is broad at the top, your shoulders and arms are positioned nicely out to the sides so your arms can swing freely. With a funnel-shaped chest and shoulders positioned high and close together, you'd have to hold your arms out to get them swinging without banging into the sides of your lower chest.

As well as allowing our arms to swing freely at our sides, there's another reason why the human chest may have become barrel-shaped. Your ribcage is joined to your pelvis by muscles – three layers of muscle at the sides, curving around onto the front, and a pair of long, strap-like rectus abdominis muscles at the front, flanking the midline. These muscles form the walls of your belly, but they also help to move your trunk. The muscles at the sides produce side-bending and twisting movements, while the rectus muscles flex your trunk forwards – these are the muscles you use to do sit-ups. The shape of the bottom of your ribcage, forming the upper attachment for these muscles, tends to reflect the shape of your pelvis, forming the lower attachment. And the shape of the pelvis has changed radically over the course of human evolution.

Apes have wide pelves, matched in shape by their funnel-shaped chests, which splay out wide at the bottom. In humans, with tall, narrow pelves adapted for walking around on two legs, the bottom of the ribcage has become correspondingly narrow. But the volume inside the chest is important, because of course it contains the heart and lungs. So it may be that walking on two legs and having a narrow pelvis is what really drives the change in shape of the human thorax: forced to become narrow at the bottom, the human chest also had to become wider in the middle and at the top, creating that characteristic barrel shape, in order to maintain enough lung volume.

Whatever it was that drove the change in chest shape, it seems that an ape-like, funnel-shaped chest, with high, 'shrugged' shoulders, reflects a lifestyle that involves a lot of climbing in trees. On the other hand, humans, who habitually walk around on two legs, have low, wide shoulders attached to barrel-shaped chests. This means that looking at chest shape could be useful if we're trying to work out how ancient hominins moved around, and might even help to reveal when our ancestors became fully committed to walking on the ground. If we assume that the last common ancestor of humans and chimpanzees had a conical thorax, the change to a barrel-shaped, human-like chest would be a real milestone in human evolution.

FOSSIL RIBS

Unfortunately, ribs are thin bones and easily broken. Well-preserved fossil ribs are few and far between, so it's difficult to track how chest shape has changed over time. But there are a few fossils that provide us with important clues.

The earliest fossil evidence of a really human-like, barrel-shaped thorax comes from an early fossil of *Homo erectus*, dating to around 1.5 million years ago. This thoracic skeleton belonged to a remarkable young man who is now known as Nariokotome Boy, and whom we'll meet properly in the next chapter.

Going back before *Homo erectus*, fossilised ribs are few and far between. *Homo erectus* isn't the earliest species to be graced with our own genus name, *Homo*, there are two earlier human species, which first appeared around 2.4 million years ago: *Homo habilis* and *Homo rudolfensis* (although this is a tricky subject, and not everyone agrees that these two are human enough to be called *Homo*). These two characters are known mostly from skulls, mandibles and teeth, with very little evidence from the rest of the skeleton.

There are a few fossil ribs from a species which has been suggested to be ancestral to *Homo*. This species was discovered by the nine-year-old son of a palaeontologist in South Africa, in 2008. While his dad was working on a dig nearby, the boy stumbled across a rock with a bit of what looked like bone sticking out of it. He dutifully took the rock to his dad, who recognised the protruding bone as a clavicle, or collarbone, and turned the rock over to reveal, to his amazement, part of a jaw with a tooth still attached to it. The little boy, Matthew Berger, had found the first fossils of an entirely new species which his dad, Lee Berger, would name *Australopithecus sediba*. The bones come from a pit in a cave, which has been described as a 'death-trap' into which the hominins fell. In the southern African Sotho language, 'sediba' means 'well' or 'spring'. Since that first chance discovery, the site at Malapa has yielded hundreds of bones from several individuals of this new australopithecine species, dating to around two million years ago.

Australopithecus sediba has a strange collection of features, some of them more archaic-looking, others more human-like. The Malapa fossils include a good collection of rib fragments, mostly from the upper and middle parts of the chest, and the top of this hominin's ribcage looks quite 'ape-like' – it's conical. There's no doubt that *Australopithecus sediba* is a hominin and walked on two legs, but if its whole ribcage was funnel-shaped, it wouldn't have been able to employ the energy-saving arm-swinging that we do while walking. This suggests that these hominins wouldn't have been able to walk or run for long distances particularly efficiently.

Looking at Nariokotome Boy and *Australopithecus sediba*, it's easy to assume that early hominins had funnel-shaped chests and later ones developed barrel-shaped chests, like ours. But another fossil calls this simple trajectory into question. In 2005, the fossil of Kadanuumuu was discovered in the Afar region of northern Ethiopia – the same broad area which yielded the famous fossil of *Australopithecus afarensis* known as 'Lucy'. Kadanuumuu probably belongs to the same species as Lucy, but he's much bigger than her; in fact, his name means 'Big Man' in the local Afar language. While Lucy would have been about a metre tall (just for comparison, that's about the average height of a three-year-old), Kadanuumuu would have been about 1.5–1.7 metres tall.

Kadanuumuu has five fairly complete ribs. It's enough to suggest that his chest was wide at the top – so more barrel-shaped than funnel-shaped. Along with other anatomical clues, Kadanuumuu's ribs have been used to suggest that he had completely abandoned life in the trees and was committed to walking on the ground.

This seems odd. If Kadanuumuu and *Australopithecus sediba* are both direct ancestors of humans, this means that early hominins probably started off with an 'ape-like' thorax, then developed a 'human-like' thorax (Kadanuumuu), before reverting to something a bit more 'ape-like' (*Australopithecus sediba*) and finally returning to a 'human-like' pattern again by the time of *Homo erectus*. It makes it look as though *Australopithecus sediba* made a retrograde step, reverting to having shrugged shoulders and presumably going back to spending quite a bit of time hanging around in trees, when it wasn't walking on two legs on the ground.

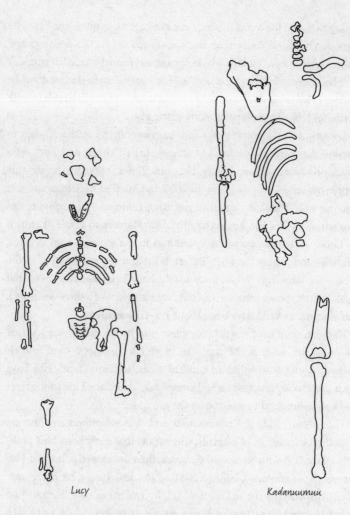

Lucy *Kadanuumuu*

Two famous fossils of Australopithecus afarensis

But this is an unhelpful way to think about evolution, for a number of reasons. We shouldn't necessarily expect simple trajectories and trends. Hominin evolution, like the evolution of any closely related group of animals, isn't a linear 'march of progress'. Rather than one twig growing out from a branch on the vast tree of life, with a neat, linear, succession of species, the six to seven million years of hominin evolution is itself a small branching tree, and each species on this cluster of twigs would have been adapted to its environment in ways we're just beginning to understand. A diversity of chest shapes among hominins may simply reflect diverse ways of moving around, with various hominins spending different amounts of time in trees or walking around on the ground.

Sometimes it's difficult to know where species belong in the hominin family tree. We can suggest connections between species but we can't really be sure – some species may form real links between the most ancient hominins and us, whereas others may be heading off on separate branches. So another possibility is that we've tried too hard to shoehorn species into a particular lineage: perhaps *Australopithecus sediba* isn't a human ancestor after all and is in fact off on a side-branch of the hominin family tree.

At the end of the day, we're trying to understand why we have got a different chest shape from that of other apes. We're trying to find out when barrel-shaped chests replaced funnel-shaped ones in our own lineage. But – and this is an intriguing possibility – maybe the assumption we're starting with is wrong. We've assumed that the earliest hominins had funnel-shaped chests, but the evidence is really too sparse and fragmentary to be sure about that. What if the earliest hominins had human-like, barrel-shaped chests right from the start?

Funnel-shaped chests are typical of non-human apes, but there is an exception. While orangutans, chimpanzees and gorillas have funnel-shaped ribcages with flaring lower ribs, the smaller-bodied and more distantly related gibbons and siamangs have barrel-shaped thoraces – rather like miniature versions of ours. Humans lack key muscle adaptations to suspensory behaviour (like hanging from arms) that the other great apes show in their anatomy. Either humans have lost these muscle adaptations in the time since an arm-hanging ape ancestor, or those adaptations (and that behaviour) never existed in our ancestry. More than 30 years ago, based on this observation about muscles, some palae-oanthropologists suggested that humans had most probably evolved from a small ape with a chest similar in shape to that of a gibbon.

This turns ideas about the evolution of chest shape on its head. We have assumed that the last common ancestor of humans and great apes would have had a funnel-shaped chest, and that the living (non-human) apes have kept that shape, while human chests have changed shape to become more like squashed barrels. But it could be that the last common ancestor had a barrel-shaped chest, and that the great apes have evolved funnel-shaped chests as an adaptation to their fondness for vertical climbing. Humans, on the other hand, might be the conservative ones in this view, holding on to a barrel-shaped chest with an ancient pedigree. We always seem to start thinking of ourselves as the special ones, the ones who must have changed most since a common ancestor with the other apes. Chest shape may be an example of where humans have actually changed less than other apes over evolutionary time.

NEANDERTHAL CHESTS

While the shape of your chest reflects your posture and how you move (as an ape with an upright trunk, which doesn't hang around in trees any more, and which tends to swing its arms as it walks), it also tells us something more general about your way of life. Or at least, the *size* of your chest tells us something. And that 'something' emerges when we compare ourselves with the closest of our closest extinct relations: Neanderthals.

The most complete Neanderthal ribcage to date comes from the Kebara 'K2' specimen – the same partial skeleton which included the Neanderthal hyoid bone we met earlier in the *Speech and Gills* chapter. Incredibly, the K2 thorax comprises fragments of every rib, and so, compared with the debates over the few bits and pieces of ribcages of earlier hominins, this leaves much less room for doubt about the shape of a Neanderthal chest.

Looking at Neanderthal bones, I am always struck by just how chunky and robust they are compared with modern human bones. Even the stockiest Anglo-Saxon skeleton looks weedy in comparison with a Neanderthal. Each bone in the body (apart from, rather strangely, the clavicle, which is extremely long and thin) is like a supersized version of the modern human equivalent. The shape of the whole skeleton, put together, is also sturdy. The ribcage looks massive. In comparison with ribs in modern human populations, the K2 ribs are larger – in absolute terms and relative to stature. Long ribs like these mean that the volume of the Neanderthal chest must have been larger than that of modern humans. But because even these well-preserved Neanderthal ribs have suffered some warping since becoming buried in the ground, it's difficult to be sure about whether that increase in volume resulted from a front-to-back expansion of the chest, or a widening. Someone needs to find some undistorted Neanderthal ribs to answer that question.

Nevertheless, even with slightly warped ribs, it's clear that the Neanderthal chest was not only larger, but a different shape to ours. The uppermost and lowermost ribs are a similar size to those in the chest of a modern man, but the middle ribs are much larger. If the modern human chest is classically described as barrel-shaped, then the Neanderthal chest is 'hyper-barrel-shaped' – much broader in the middle. It seems likely that a large chest would have meant large lungs (although even this is far from certain, as the position of the diaphragm and how far it domes up into the chest has a very large influence on lung volume). Having said that, among modern humans at least, a large chest does mean large lungs inside it. Indigenous people living in the high Andes today tend to have large chest circumferences, and large lungs inside those chests.

Neanderthals have often been described as being 'cold-adapted', and it's been suggested that their large chests and lungs were just such an adaptation. However, there seems good reason to believe – even in the absence of piles of pristine ribs – that the ribcages of more ancient hominins from Europe and Africa were also large. We've already noted that chest shape 'matches up' with pelvic shape. The recent discovery of a very wide *Homo erectus* pelvis (from a site called Gona in Ethiopia, which we'll revisit later) suggests that a wide, stocky frame was part of the characteristic body plan of ancient *Homo*. This means that, rather than developing a new and unusual stockiness, with a large chest, as part of an adaptation to cold climates, Neanderthals were just sticking to that primitive body plan, whereas modern humans tore up the blueprint, evolving to be much more slender and gracile than our recent ancestors. Large chests in Neanderthals and earlier species were probably necessary to allow the level of oxygen consumption needed for both a large body and a very active lifestyle – compared with humans today. Whereas we need around 2000–2500 calories a day, the daily energy needs of a Neanderthal have been estimated to be some 3500–5000 kcals per day. A high calorie intake, which is linked to higher heat production in the body, may have meant that Neanderthals would have been naturally suited to cold environments, but that's very different from suggesting that their body shape *arose* as adaptation to such climates.

THE HEART

There may be important differences in the shape of our chests compared with those of our ancestors and other living apes, but what goes on inside the chest is very much the same for any ape, indeed any mammal. Look inside the chest of any mammal and you'll find a pair of lungs and a heart with four chambers: a left atrium and ventricle, and a right atrium and ventricle. The right side of the heart pumps deoxygenated blood out to the lungs, where it picks up oxygen, before returning to the right side of the heart, which pumps out that nicely oxygenated blood around the rest of the body. The cycle can be illustrated as a figure-of-eight, looping through the right and then left side of the heart. But the two sides of the heart contract at the same time, so the right ventricle is pushing blood out into the lungs at the same time as the left ventricle is pushing blood into the aorta and out to the body. This double circulation is characteristic of air-breathing land animals, including us. Fish have a different sort of heart, with no left and right side to it, no double circulation. But your own embryological development once again reveals links back to your ancient fishy ancestors and their simpler hearts.

EMBRYONIC DEVELOPMENT OF THE HEART

Your heart started to develop incredibly early in embryonic development. Even when you were still a flat jam sandwich of a trilaminar germ disc, before you had rolled up in a basic body of nested cylinders, cells were getting together to make the precursor of a heart. These cells formed tubes in a double horseshoe shape around the edges of the germ disc. In the fourth week, as you started to curl around, the limbs of the horseshoe were brought together in the front of your forming body, and the two tubes melded into each other to form a single, primitive heart tube. As the branchial arches appeared on the sides of your embryonic neck, a series of aortic arches formed inside them, connecting this heart to two vessels lying along your back: your dorsal aortae.

Just four weeks after conception, your tubular, embryonic heart started to beat, and newly formed blood cells were pushed through newly formed blood vessels. The heart and circulation is a mass transport system. Whereas single cells and small clusters of cells can exchange gases, nutrients and waste with their environment through diffusion alone, when you grow a bit bigger, you need a transport system to carry stuff around. Oxygen and nutrients are carried to cells far from the sites of gas exchange or absorption of nutrients, and waste is carried away to sites where it can be removed from the body. The heart and circulation form just such a mass transport system.

When your embryonic heart started pumping, your circulation was operating in a single loop, similar to that of a fish. At this moment in your development, your tiny embryonic heart also looked very much like a fish heart. This is one instance where Haeckel could be forgiven for his interpretation of recapitulation, with 'higher' animals passing through embryonic stages where they look like *adults* of 'lower' animals.

A fish's heart is a tube of muscle, with blood coming in at the 'tail' end and pumped out at the 'head' end. From there, the blood flows into the arteries supplying the gills, picking up oxygen, and then passing into the two dorsal aortae, which join to form a single vessel further down the fish. The heart and the aortic arches of a five-week-old human embryo look uncannily like those of an adult fish. But the blood is getting oxygenated somewhere quite different: not in the gills (because although you have branchial arches, they will never be *actual* gills), and not in the lungs (because these are only just starting to develop, and of course they are full of amniotic fluid, not air). As an embryo, and as a fetus (officially from eight weeks after conception until you're born), you got all the nutrients and all the oxygen you needed from your mother – via the placenta. A single vein carries the oxygenated blood from the placenta to the embryo, where it quickly reaches the embryonic heart and is pumped out via the arteries in the branchial arches, into the dorsal aorta and round the body. Two umbilical arteries carry deoxygenated blood from the embryo back to the placenta.

A single-loop circulation suited you while you were a fetus, but it wouldn't work once you are born and start breathing air. A single circulation suits an adult fish, which picks up its oxygen from water and is surrounded by water. The fish's heart can pump out blood with sufficient pressure to pick up oxygen from the gills *and then* travel around the rest of the body. As air-breathing creatures, we exist in a much lower-pressure environment than a fish, and the pressure within our lungs drops even lower than the atmospheric pressure around us as we expand our chests to draw air into our lungs. If you had a single circulation which would push blood out under enough pressure to pass through the lungs and then around the rest of your body, the blood pressure in the lungs would be too high: it would push fluid out of those thin-walled capillaries into the air spaces of the lungs.

So this is the reason for a double circulation: one side of the heart pumps out blood into the lungs at relatively low pressure, where it can pick up oxygen and then return to the other side of the heart. This time, the blood is pumped out at much higher pressure, enough to get it right up to your head and out to the tips of your fingers and toes, propelling it out to all the capillaries in all the peripheral tissues of your body. In the pathological condition of pulmonary venous hypertension, where the left side of the heart fails to pump blood efficiently, the blood pressure in the lungs becomes too high. The double circulation then fails to do its job of allowing blood to be pushed through the lungs without fluid being driven out into the air sacs. Untreated pulmonary venous hypertension leads to the lungs gradually filling with fluid. The air sacs where gas exchange should be happening fill up, and the patient begins to drown in their own fluid.

As an embryo, floating in your pond of amniotic fluid and getting all your oxygen from your mother via the umbilical vein, you didn't breathe air and you didn't need a low-pressure pulmonary circulation. You had a single circulation like a fish. Your heart even looked like a fish's heart to begin with. However, by the time you are born, that heart needs to have left and right atria and ventricles, but it also still needs to work within a single circulation, until you emerg into the world and take your first breath. So the embryonic development of the heart is complicated and must involve the creation of a valve – a valve which will allow blood to flow from one side of the heart to the other in the embryo, but which will shut at birth, isolating the left and right sides and the pulmonary from the systemic circulations.

So there you are, at the end of four weeks' development, with your tubular, beating heart connected back via aortic arches to the dorsal aortae. That heart has a series of chambers, including a primitive atrium and a primitive ventricle. Your heart tube begins to twist into an S-shape,

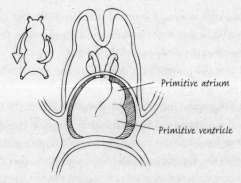

The primitive heart tube of the human embryo twists into a more compact shape

Primitive atrium

Primitive ventricle

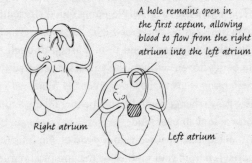

The first septum grows down from the roof of the common atrium

A hole remains open in the first septum, allowing blood to flow from the right atrium into the left atrium

Right atrium

Left atrium

A second septum grows down to the right of the first

A hole – the foramen ovale – remains in the second septum, allowing blood to flow through

The two septa, with the staggered holes in them, form a valve which allows blood to flow through in the fetus, but which will close at birth.

The embryonic heart developing

at this point looking very much like a fish's heart. But the twist keeps going, pushing the primitive atrium behind the ventricle. Then the division of the left and right sides begins, with partitions growing into the heart, like sliding doors being drawn across. The ventricle is completely divided in two, as is the outflow tract which will form the ascending aorta and the pulmonary trunk. But the partitioning of the embryonic atrium is more complex. A dividing septum grows down from the roof of the atrium, but just before it reaches the floor, it starts to perforate at the top, leaving an oval opening, or foramen, between the right and left atria. A second septum grows down to the left of the first. This one stops short of the floor, leaving a hole at its base. This cunning arrangement means that blood can flow from one side of the heart to the other in the embryo. Oxygenated blood from the umbilical vein eventually flows into the heart via the inferior vena cava, into the right atrium and flows straight through the oval foramen into the left atrium.

The fetal circulation works for the fetus, but the stage is set for a transformation at the moment when the baby takes its first breath of air. At that moment, the lungs expand and pull in air for the first time. The low pressure in the expanded lungs also pulls blood into them, and a tide of oxygenated blood rushes through pulmonary veins into the left atrium. The rise in pressure in this atrium pushes the second septum against the first, shutting the valve between the atria – they are now functionally separated. Over the coming months, the two septa will fuse together, but traces of the original openings will remain in the adult heart. If you were to look inside your right atrium, at the wall which separates it from the left atrium, you'd still see an oval depression or fossa in that wall: the remains of the oval foramen.

Embryonic development is complicated, and in some ways it's astounding that it doesn't go wrong more often than it does. The creation of this valve in the heart means that there is an opportunity for things to go awry. If the septa don't overlap, the valve can never close properly, leaving a permanent hole in the heart. A tiny hole doesn't matter too much, and many close on their own, but a large hole will mean that blood is pushed from the left to the right atrium, increasing the pressure in the right side of the heart – exactly what the principle of a double circulation is designed to avoid. High blood pressure in the

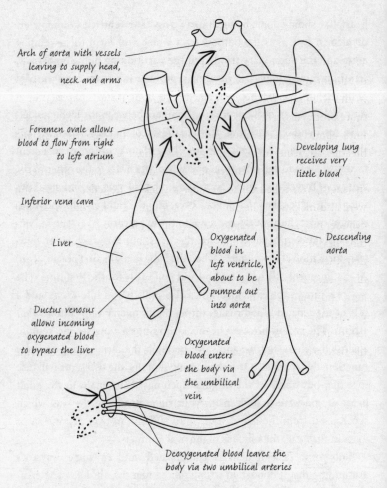

Arch of aorta with vessels leaving to supply head, neck and arms

Foramen ovale allows blood to flow from right to left atrium

Inferior vena cava

Liver

Ductus venosus allows incoming oxygenated blood to bypass the liver

Developing lung receives very little blood

Oxygenated blood in left ventricle, about to be pumped out into aorta

Descending aorta

Oxygenated blood enters the body via the umbilical vein

Deoxygenated blood leaves the body via two umbilical arteries

The fetal circulation

lungs is bad news and will eventually lead to permanent changes in the heart and blood vessels as they adapt to the high pressure. Fortunately, it's now possible to identify these holes, using a type of ultrasound scan called an echocardiogram, and to mend them surgically. Some can even be repaired using keyhole surgery.

MAKING LUNGS

Whereas the heart starts beating after just four weeks of embryonic development, and plays such an important role within the fetus while it's in the womb, the programme of development for the lungs is much more protracted. This makes sense: the lungs won't actually be needed until about 40 weeks after conception, when the baby is born and breathes air, acquiring oxygen for itself for the first time.

When, as a minute embryonic germ disc just a few millimetres in length, you rolled up to become a stack of nested cylinders, you were also making the beginnings of your guts. The narrow, innermost cylinder made of endoderm is your primitive gut tube. And it's from this tube, in your fifth week of development, that your lungs start to form. At first, there's just a tiny bump, a lung bud, growing forwards from the upper end of the gut tube. The bud grows and branches, then the right lung bud branches into three and the left divides in two – and so the pattern of your adult lungs, with three lobes on the right and two on the left, is already established. A network of blood vessels grows out from the sixth arch artery, around the growing lung buds: the precursor of the

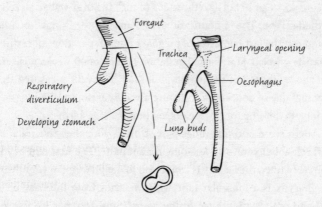

The lung buds sprout from the foregut of the embryo

pulmonary circulation. Over subsequent weeks, the buds keep growing and branching, dividing some twenty times until there are thousands of tiny twigs at the periphery of this respiratory tree. At the ends of these twigs, the beginnings of the alveoli, the bubble-like structures where gas exchange will occur, start to form. A group of these bubbles forms at the end of each twig, like a tiny blackberry.

The lungs develop in such a leisurely way that they cause problems if babies are born prematurely; if a baby is born too early, the alveoli may be too immature to allow gas exchange in the lungs. The cells lining these tiny sacs start off quite plump, but then flatten, getting progressively thinner and thinner, and capillaries in the connective tissue around the sacs proliferate. By the 24th week of gestation, there are enough flattened alveolar cells and enough capillaries around the alveoli to give the baby a fighting chance of survival.

Another important consideration in immature lungs is a lack of surfactant, an oily substance secreted by some of the alveolar cells, which helps to lower the surface tension inside the alveoli and prevent them collapsing. There's a danger of collapse without surfactant because of the way the lungs operate, by expanding to produce a negative pressure with respect to the air pressure outside the body. That negative pressure sucks air in but also threatens to suck the thin-walled alveoli in on themselves. The amount of surfactant in the lungs increases throughout the last trimester of pregnancy, in preparation for birth. In the last few weeks of a normal gestation (which is 40 weeks measured from the mother's last menstrual period, or 38 weeks measured from the actual day of conception), mature alveoli develop in the lungs and continue to develop for some eighteen months after birth.

It might seem odd that lungs bud off from your digestive tract, but just think about your own anatomy for a minute. The first part of your respiratory tract, the larynx, is connected to the pharynx in your throat. The pharynx is a muscular tube which reaches from the back of the nasal and oral cavities down to the oesophagus. In other words, your respiratory tract is *still* attached to your digestive tract.

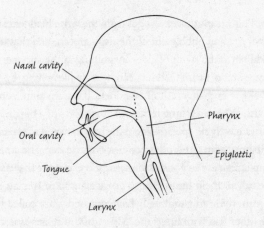

Nasal cavity

Pharynx

Oral cavity

Epiglottis

Tongue

Larynx

The larynx is a tube attached to the pharynx

Everyone who lives on land, and that's just about any tetrapod you care to mention (any amphibian, reptile, bird or mammal, that is) needs to obtain oxygen from air. Some amphibians, including the largest group of salamanders, absorb oxygen through their skin, but most of us land animals have developed another way of getting oxygen into our blood: we have lungs. It seems like a huge leap, to go from living in water using gills to extract oxygen from water, to breathing air using lungs. But a more careful look at fish anatomy makes the leap seem less daunting, and that's because air bags which form as out-pouchings from the primitive gut tube of the embryo are not exclusive to air-breathing tetrapods.

Most bony fish, from gobies to sturgeon, possess a swim bladder. This is an air bag, and it develops in the fish embryo as an out-pouching from the gut tube. Not only that, but the walls of swim bladders contain networks of small blood vessels which allow gases to diffuse out of blood into the swim bladder and back into blood again. In many fish, the original connection between the gut tube and the swim bladder is lost, but in a few species, including some carp, trout, herring and eels, the duct stays open. This means that excess gas from the swim bladder can escape into the pharynx and bubble out of the fish's mouth, and that these fish can also add air to the swim bladder by gulping at the surface. In other words, these sorts of swim bladders can operate like a lung, by allowing a little bit of gas exchange. All of a sudden, the transition from water-breathing fish to air-breathing tetrapods seems less of a hurdle.

There are a few living species where the link between water-breathing fish and air-breathing tetrapods becomes even clearer. The air bags in a select group of fish are so good at allowing gas exchange that it's entirely reasonable to call them lungs. These lungs even have tiny air sacs like ours. The fish – you might already have guessed – are called lungfish. (There are other fish with lungs too. More distant relatives of ours, the ray-finned fish known as bichirs, have lungs, but their lungs are smooth inside, not full of tiny air sacs like our lungs or those of lungfish.)

Perhaps unsurprisingly, among all the other bony fish, lungfish are closest to the tetrapods. There are three groups of lungfish alive today, in Africa, Brazil and Australia. All lungfish have gills as well as lungs, but although the Australian lungfish can depend on their gills to oxygenate their blood, the African and Brazilian fish would suffocate if you held them underwater for any length of time. They have made a commitment to breathing air which they have inherited from their ancestors, and which they share with their amphibian cousins. Like amphibians, lungfish have lungs connected to the pharynx and supplied with blood from the sixth aortic arch. This all sounds very familiar, because it's the same basic pattern as in us. Unlike other fish – but like amphibians – lungfish have a heart in which the atrium and ventricle is partially divided into two. Oxygenated blood from the lungs is kept largely separate from the deoxygenated blood returning to the heart from the rest of the body. The outflow tract from the heart contains a spiral valve which helps to keep the two streams of blood separated, sending the oxygen-poor blood into the lower pairs of aortic arches, which carry it off to the lungs and gills, and the oxygen-rich blood into higher arches, to be sent around the rest of the body.

Lungfish might show us – as a living example – how the lungs of land animals originated. Their lungs are also a fantastic example of evolution's propensity for recycling and repurposing, as existing structures (and genes) are coopted for new functions. Sometimes, existing structures and genes get duplicated before new roles can be taken on. In our DNA, gene duplications free up genes to take on new roles, but a similar process can act with fairly major anatomical structures, such as the duplicate jaw joint in early mammals, which allowed the bones around the original joint to migrate into the ear to become auditory ossicles. Another way in which structures can take on new roles is if they serendipitously happen to fulfil a new role, just as an old one is becoming redundant. An air bag attached to the gut, which was *already there* in our water-breathing, fishy ancestors, proved essential for air-breathing when our tetrapod forebears hauled themselves out onto land. Evolution loves to recycle, to make the most of unforeseen opportunities.

GUTS AND YOLK SACS

Links with egg-laying ancestors

and fruit-eating apes

'We admit that we are like apes, but we seldom realize that we are apes'
RICHARD DAWKINS

THE GROWTH AND CONVOLUTIONS OF
THE GUT TUBE

Before interesting offshoots like swim bladders or lungs start to bud off, an embryo's gut – whether you're talking fish, fowl or human – is a simple tube of endoderm: the innermost layer in that rolled-up body of nested cylinders. The tube is blind-ended to start with, as there's no opening at either the mouth or the anus. In the middle, the tube is connected to the yolk sac which lies outside the body of the embryo.

The existence of a yolk sac in a human embryo is another evolutionary echo: a reminder that you came from ancestors which laid eggs. In embryos that develop inside eggs, having to fend for themselves outside the mother's body, the yolk sac is large and is an important source of nutrients. The yolk of a hen's egg is massive compared with the size of the early chick embryo, because it needs to sustain the developing chick up until the point it hatches and starts to find its own food. In placental mammals (that includes us) where embryos stay inside their mother's body for longer, the placenta will provide the embryo with nutrients, and oxygen, as well as removing its waste. Here, the yolk sac is effectively obsolete, but it still develops. The yolk sac of an early human embryo is very small compared with the yolk of a bird embryo and it eventually degenerates until, once the baby is born, it is nowhere to be seen.

The yolk sac develops from the original cavity inside the blastocyst – the hollow ball of cells which develops during the first week of gestation and implants into the wall of the uterus. When the embryo is at its jam-sandwich, trilaminar-disc phase, the yolk sac lies next to the lower, endoderm layer of the disc. In fact, endoderm cells also spread around the inside of the yolk sac to line it. Then the embryonic germ disc rolls itself up into those nested cylinders and the endoderm forms the innermost cylinder: the gut tube. The connection to the yolk sac is still there, eventually becoming a narrow duct called the yolk sac stalk.

The gut tube is described as having three parts: the foregut, towards the head end; the midgut, which is where the yolk sac stalk attaches; and the hindgut, towards the tail end of the embryo. The foregut is destined to form the cavity of the mouth, the pharynx – including the bud which will form the larynx, trachea and lungs, the oesophagus and the stomach.

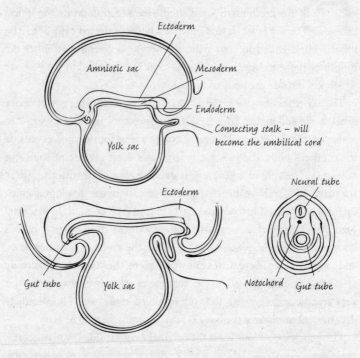

The upper diagram shows a slice through the long axis of the embryo at an early stage of folding. The lower diagram on the left shows a similar slice at a later stage, and the diagram on the lower right shows a cross section through the embryo at the same stage

The midgut lengthens enormously to form the coils of the small intestine and half of the large intestine, while the hindgut forms the rest of the large intestine, right down to the rectum and anal canal, and a bit of it also gets pinched off to form the urinary bladder.

Your stomach started to appear in the fifth week of your embryonic development, as a swelling in the foregut tube, then it began to grow like a banana, with one side growing faster than the other, pushing it into a curved shape. Further growth transforms the stomach into a roomy bag, bearing a closer resemblance to its adult form when, as well as digesting food, it will store it, releasing just a little at a time for further digestion in the intestines. Accessory organs of digestion bud off from the foregut: the liver, gallbladder and the pancreas begin as sprouts from the gut tube. These organs always retain their original links with the gut tube: the connections become the ducts through which their secretions (bile from the liver and digestive enzymes from the pancreas) pass into the small intestine.

Beyond the point where the liver and pancreatic buds appear, the midgut elongates. It does this so quickly that the embryo's body can't

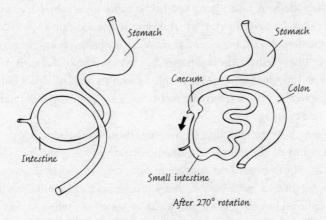

After 270° rotation

The midgut lengthens and rotates

keep up, and the gut outgrows the embryonic abdominal cavity and pushes out into the umbilical cord. In other words, the gut herniates. So if anyone asks you if you've ever had a hernia and you would normally have said no, then the answer, for everyone, is yes. Everybody has had this hernia, where their embryonic intestines pushed out of the front of their abdominal wall, starting in the fifth week of development. Fortunately, in the vast majority of cases, the external coil of intestine returns to the abdominal cavity as the embryo continues to grow, past eight weeks of development, when it becomes known as a fetus. As the loop of intestine pushes out and then tucks itself back inside the body, it undergoes a rotation, so that the lower limb of the loop ends up lying in front of the upper limb. This creates an arrangement which is present in your abdomen as an adult: the transverse colon, part of the large intestine which lies horizontally in the abdominal cavity, lies in front of the very first part of the small intestine, the duodenum.

As with any sequence of events in embryology, things can go awry with gut development. By twelve weeks of development, all of the intestines should have returned to the abdominal cavity, but occasionally (in about four babies in every 10,000 born), the herniated gut fails to recede, leaving the fetus with a persistent hernia called an omphalocele (from the Greek for navel and cavity). This type of defect is usually picked up on the twenty-week ultrasound scan. When the baby is born, the hernia looks like a translucent bag protruding into the umbilical cord, with coils of intestine visible inside it. A baby with an omphalocele requires corrective surgery to return the protruding guts to the abdomen and to seal the opening.

Sometimes the intestines successfully return to the abdominal cavity but fail to do the normal 270-degree rotation to bring the transverse colon to lie in front of the duodenum. Such cases of malrotation can be inconsequential and asymptomatic, but others can cause problems like an unhelpful twisting of the gut, which closes off the tube and cuts off blood supply.

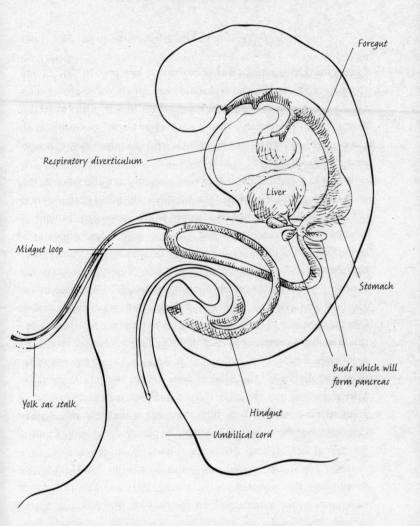

The developing gut tube at the beginning of the sixth week of development, with the swelling stomach, the liver and pancreas budding off, and the midgut loop

A FANTASTIC VOYAGE

I know that I have a minor malrotation in my own guts. In 2007, I tentatively agreed to undergo an exploratory capsule endoscopy, more colloquially referred to as a 'pillcam'. I went along to Selly Oak Hospital in Birmingham, where gastroenterologist Dr Roy Cockel was ready to help me visualise my own intestines, with the help of a tiny camera contained in a yellow capsule.

When Roy and his colleagues first started using the pillcam, they were able to visualise regions of the intestines which had effectively been off limits before. Tube-like endoscopes which are passed through the mouth, down the oesophagus and into the stomach, can only see as far as the first part of the small intestine, the duodenum (this type of investigation has the wonderfully elongate name of oesophagogastroduodenoscopy. Luckily, it's usually abbreviated to OGD). Endoscopes can also be introduced at the other end of the gut tube, via the anal canal, up into the rectum and colon and as far as the end of the small intestine – this is called a colonoscopy. But the huge length of small intestine – metres and metres of it – between the duodenum and the end of the ileum remain beyond the range of these flexible cameras. Using CT or MRI scans doctors can see the coils of small intestine and put contrast medium into them so that the lumen or space inside the intestine can be visualised, but they have not been able to peer in detail at the lining of this part of the gut. Until, that is, the pillcam was invented.

When Roy Cockel started using pillcams at Selly Oak Hospital, the *Birmingham Post* reported on the advancement and likened it to the experience of the submarine crew in the 1966 film *Fantastic Voyage* (where a team of doctors are put into a submarine, miniaturised and injected into a patient's bloodstream to get to the brain to destroy a clot). Capsule endoscopy is perhaps a little less glamorous and a little more controlled (no submarines threatening to return to normal size in this version), but it provided a new opportunity to see the lining of the entire small intestine, which would prove hugely useful in patients suffering from intestinal bleeding, because now the source of such bleeding could be pinpointed.

When I met Dr Roy Cockel for my own pillcam investigation, he started the procedure by applying small electrodes to the surface of my belly before strapping on the data recorder, which would track and record the images taken by the pillcam. Then he showed me the pillcam itself. It was somewhat larger than I'd anticipated, but nevertheless, not much bigger than a large antibiotic or vitamin capsule. As he held the pillcam between finger and thumb, I could see it flashing away (of course, flash photography is pretty essential in the darkness of the bowels), twice a second. It would continue flashing, and taking pictures, two a second, all the way through my intestines, until it racked up 50,000 images. We could watch the images it was capturing, live, on a laptop. There were images of my open mouth, then in it went and, with a gulp of water, I swallowed it down easily. After a few minutes, the pillcam had already passed out of my stomach into my duodenum – the first part of my small intestine. I hadn't eaten anything or drunk anything other than plain water for almost 24 hours, and I was pleased to see that the pictures were beautifully clear. The lining of my intestine looked rosily pink, and slightly fluffy – the tiny fingerlike projections from the mucosa were responsible for that velvety appearance.

In my intestine, the only abnormality of the lining was what could only be described as a red pimple. It didn't look worrying, but what was more interesting to me was the path of the pillcam through my guts. Its journey recorded the twists and turns of my intestines, and when I looked at the trace of its progress, I noticed something unusual. My duodenum is different from most people's; yours probably curves around in a nice C-shape and then tucks itself up behind the stomach before turning forwards to become the second part of the small intestine, the jejunum. Mine doesn't. My duodenum, as mapped by pillcam, continues south, and the coils of my jejunum start off low down on the right side of my abdomen, whereas yours (probably) lie in the middle and upper part of your belly.

The pillcam took three hours to travel through my small intestine, helped along by the regular contractions in the muscle wall of the gut which normally push food through. The picture became cloudy as the pillcam passed into the large intestine – I hadn't starved myself for long enough to empty this stretch of my guts.

Within 24 hours, as predicted, the capsule had emerged from my body, and because I was rather in awe of this little camera which had travelled on its own fantastic voyage, I retrieved it (with the help of plastic gloves, of course), cleaned it up and disinfected it. It sat on the windowsill in my bedroom for some time, a souvenir of my intestinal photographic journey, until my husband decided that this really was too much, even for him. (I've still got it, though, in a box. I feel oddly attached to it and cannot bear to simply throw it away.)

ON THE 'UN-UNIQUENESS' OF HUMAN GUTS

On a basic level, human guts are not much different from those of most mammals which eat a generalist's diet, omnivorously consuming anything from meat to plants. I remember dissecting a dog's abdomen for the first time, at the Anatomy Department in Bristol University, where I sometimes branched out to teach vet students as well as medics. I was struck by just how similar these guts were to the human ones I was more familiar with. The dog's intestines were relatively short compared with a human's, but otherwise very similar indeed. It tends to be the strict vegetarians of the mammal world who have the most lengthy, complicated and convoluted guts. They need spaces – great fermentation vats – where they can brew up a heady mix of chewed leaves and bacteria, to get the most out of their tough diet. Some herbivores have fermentation vats located in their foreguts, like the great, multi-chambered stomachs of cows. Others wait until later in the intestinal journey to make their brew. Horses have vast caecums – a huge cigar-shaped bag attached to the base of the large intestine in the lower abdomen which reaches right up to under the sternum. We humans have neither a multi-chambered stomach nor a massive caecum; in fact, the thing called a 'caecum' in your abdomen is just the start of the colon, and not truly equivalent to the caecum of other animals. The equivalent or homologous structure in us is the appendix, which is just a narrow tube containing immune cells, but far too small to do anything useful in the way of fermenting vegetable matter. Our guts are those of a generalist, inherited from ape ancestors whose diets contained something a little more nutritious than just leaves: fruit.

But while human guts might be similar to those of any fruit-eating ape, it has long been thought that there is an important difference which may have been incredibly important in human evolution. This difference lies not in any particular specialisation in the guts, but relates to their relative size. It seems that humans have small guts for their body size. In the 1990s, this observation was linked to the fact that humans have very big brains, and the 'Expensive Tissue Hypothesis' was born.

There's no doubt that the size of the human brain is something rather special, and that our large brains somehow underpin our success as a species. All our special abilities, which include the extreme degree to which we copy and cooperate with each other and create culture, must somehow depend on that enormous organ in our heads. And yet this organ is very demanding. Brains are extremely energy-hungry, consuming around 20 per cent of our entire daily energy requirements while making up for only 2 per cent of our body mass. It seems that our ancestors must have found a way of paying for this 'expensive tissue' inside our skulls – either by boosting daily energy intake or by compensating by making savings elsewhere.

If our ancestors, and indeed you or I today, paid for our brains simply by obtaining more energy, our bodies would have to deal with a high turnover of energy, and there should be an obvious hike in the metabolic rate of humans compared with other animals. An average human, weighing some 65kg, has a 1.3kg brain – and that's one kilogram heavier than expected for a mammal of equivalent size. While the average metabolic rate of your body comes out at 1W per kg, brain tissue is much more demanding, requiring 11W per kg. So an extra kilogram of brain should cost you at least an extra 10W. That should show up in your metabolic rate, but there's absolutely no indication that our metabolic rate is any higher than would be expected for a mammal of our size.

In the mid-1990s, physical anthropologists Leslie Aiello and Christopher Dean both spotted this conundrum and believed they had solved it. They suggested that the growth of the energy-hungry brain in hominins could have been offset by a reduction in some other 'expensive tissue'. When they looked at studies of the composition of the human body, in terms of the relative sizes of various organs, they believed that they had found the answer. While organs like the heart, liver and kidneys were the size you'd expect for a 65kg primate, the guts seemed to be a lot smaller. Aiello and Dean suggested that there had been a 'coevolution' between brain size and gut size, so that as the brain grew larger, the guts shrank, while the metabolic rate stayed about the same. They went further, suggesting that the reduction in the mass of human guts was linked to a change in diet, as early hominins began to eat meat. One of the main differences between a herbivore's guts and those of a carnivore, after all, is that the latter tend to have relatively short guts. You don't need such a long intestine if you're consuming an easily digestible, high-quality diet. Later hominins could have made another energy saving by moving some of the work of digestion to outside the body – cooking food before consuming it. The energy saved by having a shorter gut could then be put to good use in growing a large brain.

Aiello and Dean weren't the first researchers to notice a link between diet and brain size, but previous authors had focused on the implications of a large brain for finding food, rather than looking at the energy demands of brains and guts. Animals eating a poor-quality, leaf-based diet need very long guts; omnivores' intestines are short in comparison, and carnivore guts are even shorter. As leaves are generally easier to outwit than more animate prey, carnivores tend to have larger brains than folivores. This means that you'd expect small guts to be associated with large brains, and vice versa. But Aiello and Dean thought that there was something more to it: that a high-energy diet and therefore small guts were *necessary* for brains to grow larger.

There were plenty of criticisms of the hypothesis at the time. Some commentators were concerned to point out that the hypothesis didn't explain *why* hominin brains expanded, merely *how* they might have been able to do so, while burning up all that extra energy. Fair enough, but Aiello and Dean had never suggested that they'd discovered what must surely be the holy grail of palaeoanthropology – the reason behind our greatly enlarged brains. Reduced guts were perhaps just a *conditio sine qua non*, without which brain expansion wouldn't have been able to get off the ground.

The hypothesis gained weight when researchers found that a rule linking small guts to large brains, and vice versa, seemed to extend across many primate species. The same seems to be true for at least a few fish. The problem is that correlation doesn't mean causation. It's still not clear that you *need* small guts in order to have a large brain. Once again, it may be that large brains tend to be associated with trickier-to-acquire, high-quality diets, and those diets can be adequately digested by relatively short guts, but that doesn't mean you *need* a short gut in order to free up energy for the brain. In fact, bats buck the trend, showing large brains associated with large guts, and small brains with small guts. Again, this seems to be related to the brain's role in tracking down food. Bats with large guts tend to be fruit-eaters, but they have large brains because the areas involved with vision and smell have necessarily expanded in order to help them find the fruit they love. The relationship between brains and guts is far from simple, and some researchers have opined that brain size and gut size vary so widely across different animals that they're as good as independent from each other.

There were other concerns about the expensive tissue hypothesis, right from the time of publication. It was suggested that the energy savings needed to support brain expansion could have been made elsewhere, by reducing body mass overall, by being a little less active, or by sleeping a bit more. And perhaps it's more evolutionarily plausible to imagine that large brains came about before downsizing guts: an expanded brain may have enabled the development of skills (like hunting and cooking) which in turn allowed our ancestors to get hold of a higher-quality diet, and the gut may have shrunk *as a result*. Indeed, there's some evidence from animals that gut size can change during the lifetime of an animal if its diet changes.

There were even doubts about whether guts really *were* small in humans. The relative sizes of organs in humans had been worked out based on a very small sample, and there are wide ranges of anatomical variation in living humans. And then there were problems with how humans had been compared with other primates, to come up with a predicted size for human guts. Most primates can be classified, extremely crudely, as either fruit-eaters or leaf-eaters. Apes tend to be fruit-eaters, and as such have smaller guts than leaf-eaters, at least using a measurement of the surface area of the guts. So if you lump all primates together to try to predict how large the guts should be in primate X, and primate X happens to be a fruit-eater (or, for that matter, an omnivore), you'll end up with an overestimate. When you plot a graph of gut area against body size, humans fall nicely on the fruit-eating/omnivore axis – just where you'd expect an ape to sit.

So where does this leave us? Certainly, there seems to be some doubt about whether our guts really are shorter than would be expected for any fruit-eating or omnivorous ape of about our size. But if Aiello and Dean's hypothesis isn't right, then how do we pay for our big brains, without increasing our metabolic rate?

A recent study, published in 2011, looked at a large bank of new data, taking in 191 specimens, representing 100 mammal species. Firstly, the study failed to find any general support for the expensive tissue hypothesis: there was no association between the sizes of brains and any other other 'expensive' organs. But the researchers found that there was something interesting going on with another, less expensive, type of body tissue: brain size seemed to be negatively correlated with body fat, in other words, fat animals tended to have small brains and vice versa. They proposed that this correlation happened because animals tend to rely on either being smart *or* having decent energy stores as ways of avoiding starvation. They also suggested that the balance between brains and fat could be due to an energy trade-off. Although adipose tissue itself has low energy demands, it could end up being expensive for the animal because of the extra weight being carried around. Finally, the researchers thought they might have the solution to the human riddle of big brains and very average metabolic rates. Humans seem to buck the mammal trend by having big brains *and* being relatively fat, compared with other apes, and it looks like the extra (low-cost) fat could be masking the effects of investing in a great big, expensive brain. If you just look at fat-free body mass, then human basal metabolic rate *is* high, compared with other apes. When you add the fat on, the average metabolic rate of the whole body drops. So the problem no-higher-than-predicted metabolic rate of humans, despite our massive brains, seems to have been been solved, or at least obscured, by fat. We still need to account for the extra energy demanded by the brain, but there's now no need to invoke a trade-off in some other energy-hungry tissue. The extra energy might have come from some savings in the function of the body, perhaps in locomotion or reproduction.

As we've already mentioned, our ancestors are also likely to have got an energy boost from their diets, which could have depended on reducing the cost of digestion through food processing – including pounding and grinding food as well as cooking it. It's likely that they also switched to eating more high-quality foods, which could certainly be meat, but we shouldn't ignore the importance of starchy foods like tubers in our ancestors' diets, too. Switching to high-quality foods, eating a flexible diet and sharing food meant that hominins had a high-energy diet that they could rely on. It may be that this was crucial, not only for that disproportionate increase in brain size, but also for an increase in body size.

But when we look at our own guts, perhaps we're not that special at all. I don't think it's clear that human guts are much different from those of our nearest ape relatives. They're bog-standard frugivore guts, but that means they're also good for an omnivore. In other words, we're pretty unspecialised when it comes to our digestive tract, but this gives us a great advantage, it means we can be very flexible in our diets. We can survive by eating a great range of different foods, and some of us alive today are almost exclusively carnivorous while others are strictly vegetarian. Looking back into our distant past, the flexibility our ancestors had by dint of their unspecialised guts meant that they would have been able to survive by adapting their diets to what was around. Fundamentally, humans are not fussy eaters (something I must remind my three-year-old daughter of).

There's some evidence that our digestive systems *have* changed, not anatomically, but in terms of the enzymes we use to digest our food. Compared with chimpanzees, we have multiple copies of the gene encoding the enzyme amylase, which breaks down starch, making us better at digesting starchy foods like tubers. In the last 10,000 years, in many different human populations, across Europe, India, Africa and the Middle East, genetic changes have meant that one gene which is normally 'switched off' during childhood stays 'switched on'. This gene codes for the enzyme lactase, which allows us to digest the milk protein lactose – something that's essential for any mammal which depends on its mother's milk as an infant. In most mammals, and in many humans, the ability to produce lactase disappears after the infant is weaned, because it's simply not necessary any more. But in populations that have been herding animals and drinking milk since prehistoric times, it's common to find the ability to produce lactase persisting into adulthood. This change *must* have happened since the domestication of animals in the Neolithic, which happened at different times in different places. Where these genetic changes have spread right across a population, it means that natural selection has acted very strongly on the ability to keep making lactase. This probably happened during drought or famine conditions, when being able to drink and digest fresh milk could have made the difference between life and death. As an aside, these recent changes in our digestive physiology mean that the whole idea of a 'paleodiet', purportedly healthier because it recreates the diet we evolved to eat as hunter-gatherers, is off the mark. Firstly, diets would have varied from place to place, even back in the Palaeolithic, and many populations have changed genetically since we left that way of life behind. We are not fossilised hunter-gatherers.

Tinkering with amylase and lactase genes aside, our digestive tracts are still pretty much standard ape guts. It seems to be the case that we've been rather too keen, in the past, to identify bits of ourselves as unique. While no one is denying that the whole human package is unique (but then of course you could say the same for any species, so don't feel too smug), the parts that make up that package are less remarkable. Things which we once thought to be uniquely human often turn out to be shared by other animals, especially our close relatives. Human guts might be one example, and perhaps we just need to come to terms with the fact that we have unremarkable ape intestines.

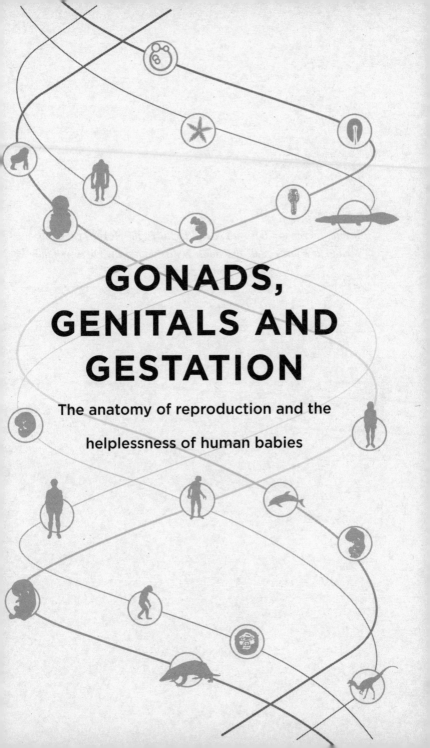

GONADS, GENITALS AND GESTATION

The anatomy of reproduction and the

helplessness of human babies

'If you . . . even touch it with your little finger, the pleasure causes their seed to flow forth in all directions, swifter than the wind, even if they don't want it to'
REALDO COLOMBO, 1559 (SELF-PROFESSED DISCOVERER OF THE CLITORIS)

LUMPS, BUMPS AND TUBES

As we follow the digestive tract to its end, travelling through those coils of small intestine and the great loop of colon lying around the edges of the abdomen, we find ourselves right down in the pelvis, where the tract finishes at the rectum and, finally, the anal canal. The pelvis is also home to the urinary bladder, which sits right at the front just behind the pubic symphysis – the joint between the two pubic bones at the front. But of course there are other organs down here as well: the reproductive organs.

The human reproductive system allows us to achieve internal fertilisation, in other words, letting sperm encounter an egg *inside a body*. It might not sound that special, but it's something that only a subset of vertebrates actually do. Most fish and amphibians do it outside their bodies: they have external fertilisation. A female fish just lays her eggs straight into water, and the male fish squirts sperm at them – done! Making the transition from water to land was a huge challenge for our ancestors in many ways: they needed to invent lungs to breathe air instead of water, they needed limbs to move around on rather than fins, and they needed a new way of sensing vibrations in the air – so they developed ears. But they also needed to tackle that sticky problem of fertilisation. If a female land animal laid her eggs on the ground and a male dumped his sperm on them, the whole thing would be a sticky mess which would dry out and die before anything interesting had a chance to happen.

One solution to this problem is to return to water to breed, which is exactly what we presume the earliest land animals did, just like most amphibians today. Frogs and toads may be happy spending time away from water, living up trees, under stones and in walls, but when the reproductive urge takes hold, they have to go back to their watery origins and find a bit of water to deposit their eggs and sperm in. Fertilisation for these creatures is external, just as it is in most fish. But some amphibians have developed a way of achieving internal fertilisation, which sounds more exciting than it actually is. In most newts and salamanders, the female keeps her eggs inside her and the sperm are introduced to the eggs within the female's body, but these amphibians achieve fertilisation without direct contact between male and female. The male 'lays' a sticky mass of sperm which is protected with lashings of mucus, and makes sure he does this close to a female. If he's lucky, the female salamander wanders over and sits on the gelatinous spermatophore. It all seems deeply unsatisfying, but salamanders must enjoy it – at least, they keep doing it (which is the key to survival of a species, after all).

From what we can observe in their living descendants, early reptiles seem to have experimented with a new way of doing things: once again, the female keeps her eggs inside, but this time the male introduces his sperm directly into the female, rather than just leaving a sticky package on the ground for her to engage with. From this point on, the stage was set for male land animals to develop a panoply of interesting delivery devices, although it must be said, these are very basic in most reptiles. In reptiles, the lower end of the embryonic gut tube becomes something called the 'cloaca'. This is a good name for it: it's connected to both the bladder and the end of the large intestine, so urine and faeces collect here before being expelled from the body, and in Latin, cloaca means 'sewer'. In a male reptile, the spermatic ducts also empty into the cloaca. In a female reptile, the large intestine, bladder and the two oviducts all open into the cloaca. Transfer of sperm in most reptiles is essentially achieved by the male and female pressing their bottoms against each other. In order to help things along, the male is able to turn two pockets inside his cloaca (called 'hemipenes') inside out, to push the sperm in the right direction.

Birds and mammals, both descended from reptile stock, also engage in internal fertilisation, involving intimate contact between the prospective mother and father. Most male birds lack a penis, so like many reptiles, they rely on pressing their bottoms up against their mate's, to encourage the sperm to pass from one cloaca to the other. But a few birds (and in fact, some reptiles too, including crocodiles and turtles) do possess an erectile penis which can be introduced into the female's cloaca to deposit sperm there. Ducks have a penis with a spiral, helter-skelter channel on the outside along which semen flows into the female. But the penis of those few reptiles and birds is something quite different from the penis in most mammals. In the male reptiles and birds who possess one, the penis arises from the floor of the cloaca. When the spongy tissue inside the penis fills with blood, the organ pops out from the cloaca, ready for work. In mammals, the penis originates on the outside of the body, and the cloaca gets completely divided up into a pouch at the front, which is called the urogenital sinus (which itself will form the bladder and the upper part of the urethra) and a back portion: the rectum and anal canal. In a male mammal, the penis forms from various lumps and bumps situated in front and to the sides of the urogenital opening in the embryo. And although the end result looks very different, the same lumps and bumps are there in a female embryo, too.

The cloacal membrane is the area where the end of your gut tube abutted the outside of your body. In the following week, as the cloaca on the inside becomes divided by the urorectal septum, the cloacal membrane on the outside also becomes divided in two – into a urogenital membrane and an anal membrane. In the same week, the urogenital membrane perforates so that the bag-like sinus above it can drain into the amniotic cavity. But the picture is still the same in both male and female embryos, up until the ninth week of development. This is a unisex kit of parts which can be used to build the genitals of either sex.

From then on, the development of your external genitalia depends on whether or not you possess a Y chromosome, which will prompt your gonads to develop into testes, which in turn produce testosterone. Without that masculinising hormone, the unisex kit of parts develops into a female

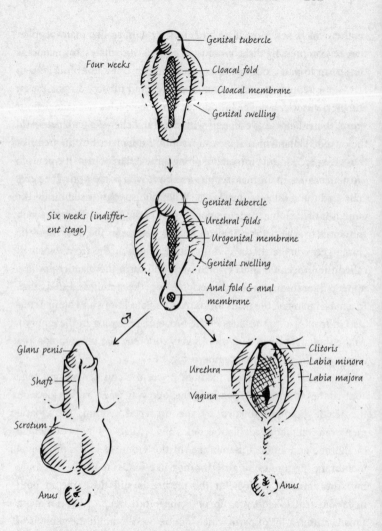

A unisex kit of parts later develops into male or female external genitals pattern: the genital tubercle becomes the clitoris, the urethral folds become the labia minora, and the genital swellings become the labia majora, as well as enlarging and merging in front of the clitoris as the mons pubis.

In a male fetus, under the influence of testosterone, the genital tubercle grows to form the shaft and tip of the penis. The outer, genital swellings enlarge to become the scrotum, and the inner, urethral folds come together and fuse, zipping up from the back to the front, to enclose the urethra inside the root of the penis. As with any aspect of embryological development, things can go wrong with the formation of the genitals, and the most common problem is a failure of the urethral folds to fuse, leading to a condition called hypospadias. This affects a few baby boys out of every 1000. The urethra may open underneath the penis, rather than at its tip, or way back, in the perineum, level with the scrotum. Surgical repair of hypospadias is very successful: it's possible to surgically reconstruct the urethra, enclosing it within the penis.

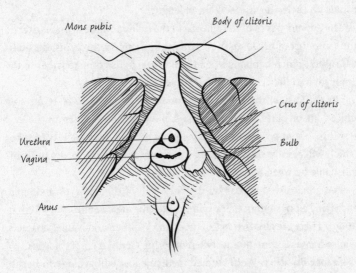

The clitoris as seen on an MRI scan of a woman's pelvis (redrawn from O'Connell et al., 2005)

Although the external reproductive organs in women are strikingly different from those of men, there are deep similarities that go right back to those early weeks of development. Under the surface, the clitoris isn't just the protrusion lying under the pubic bone. It's connected to two bodies of erectile tissue, lying along the bony ischiopubic ramus which frames the perineum on each side. These are the crura (literally, the legs) of the clitoris. Flanking the opening of the vagina, and also extending to the clitoris, there's another pair of erectile bodies known as the vestibular bulbs (the vestibule is the space between the labia minora). The body, crura and the bulbs of the clitoris are all made of spongy tissue – full of holes which fill with blood and swell during sexual arousal. The clitoris is much more extensive than you might suspect (and indeed, more extensive than shown in many anatomy books). The glans and body of the clitoris are 2–4cm long, and that's at rest – during sexual stimulation, the clitoris grows even longer. The vestibular bulbs are each about 3cm long, and the crura, stretching along the ischiopubic ramus on each side, are 9–11cm in length.

The spongy tissue in the clitoris, in the bulbs and the crura, is exactly the same sort of tissue that's present in the penis. In the penis, the bulbs are fused (which happens when the urethral folds come together in the embryo) and the crura are much longer.

On the inside, the internal reproductive organs also start off as a unisex kit of parts. It seems remarkable given how different the end result is, but at six weeks the developing gonads, and the sets of tubes which will resolve themselves into the reproductive tracts, are indistinguishable between female and male.

The gonads first appear on the back wall of the embryo's abdomen, each lying alongside another swelling called the mesonephros (or Wolffian body). There are also two sets of tubes: the Wolffian and Mullerian ducts, whose names commemorate two pioneering German embryologists.

Caspar Friedrich Wolff studied medicine and embryology in Berlin, and in 1759 he produced a dissertation on the 'Theory of Generation', which became a landmark in embryology. In it, he revived the

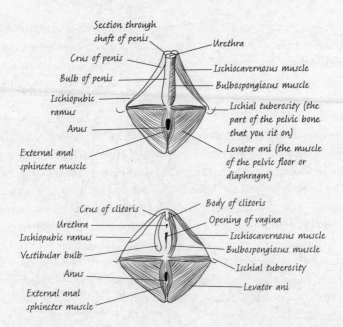

Superficially very different, male and female external genitals are similar under the skin. These illustrations show the male (top) and female (bottom) perineum: the area between the legs, with the pubic symphysis in the front and the coccyx at the back

Aristotelian idea of epigenesis, based on his own observations of plant and chick development. He wrote: 'From the theory of epigenesis we deduce that the parts of the body have not pre-existed but that they were formed gradually.' This was a contentious claim, as the prevailing theory at the time was that of preformation: the idea that embryological development involved the growth of tiny, already preformed parts. Although Wolff failed to persuade the authorities about epigenesis, he soldiered on with his embryological research, describing many aspects of development for the first time, including the formation of the gut tube.

In his original dissertation of 1759, Wolff charted the development of the chick embryo's kidneys in great detail. He described structures at the back of the chick's abdomen including the mesonephros, and the

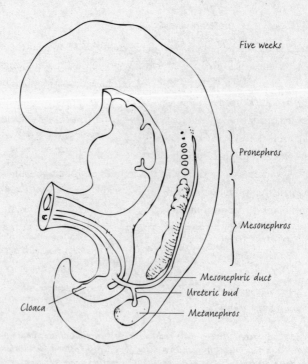

Five weeks

Pronephros

Mesonephros

Mesonephric duct

Ureteric bud

Cloaca

Metanephros

The developing embryonic kidney

Wolffian duct attaching to it. 'Nephros' means 'kidney' in Greek, and the mesonephros looks like a developing kidney, with tubules inside. It actually functions as a kidney in a chick embryo – and in a human embryo, too – filtering blood and producing urine which runs down the mesonephric duct. But while it's working in this way, a new kidney is forming: the metanephros, right down in the pelvic region of the developing embryo. Eventually, this will take over and become the definitive kidney, and it will rise up out of the pelvis to lie high up on the back wall of the abdomen. This is where your kidneys are – so high up in fact, that the top of each kidney is tucked under your twelfth rib on each side.

So what becomes of the mesonephros? In women, very little. It regresses and ends up as two small islands of tissue near the ovaries. In adult men, it's no longer anything to do with the kidneys, but the tubules inside it have been recycled to make something else: they become the sperm ducts of the testis and the mesonephric duct becomes the vas deferens, carrying sperm from the testis to the male urethra.

In chicks and humans and a lot of relations in between, the mesonephros operates as a kidney in the embryo and then gets hijacked by the male reproductive system. In more distant relatives – fish and amphibians – the mesonephros carries on functioning as the adult kidney. Once again, we're seeing that pattern of duplication freeing up structures to take on new roles. The 'invention' of the new metanephros in ancient reptiles meant that the mesonephros could stop being a kidney and do something else instead. And in fact carrying sperm was not a new role for the tubules inside the mesonephros and the mesonephric duct. Even in fish, where most of the mesonephric tubules

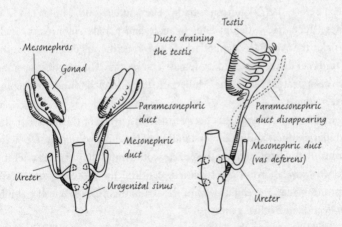

A unisex kit of parts for the internal genitals (on the left) and the pattern of development in a male human embryo (on the right), where the mesonephric (Wolffian) duct persists to become the vas deferens vas deferens

function as part of the kidney, some of them are used to carry sperm from the testis to the mesonephric duct – through the kidney. This makes sense when you look at the position of the testes and kidneys of fish, alongside each other, on the posterior wall of the abdomen (just as they are in human embryos). Again, we're seeing echoes of earlier stages of evolution in the developing human embryo, which are impossible to ignore. In this case, the echoes are there in function as well as structure, with the human mesonephros still operating as a kidney *in utero*.

In female human embryos, the Wolffian ducts disappear, and reproductive organs form from a second set of tubes: the Mullerian ducts.

Johannes Peter Müller was born in 1801 and was educated in a Jesuit seminary, destined to become a priest, but he became fascinated by biology and went on to study medicine instead, eventually becoming a professor of anatomy, physiology and pathology at the Kaiser Wilhelm Academy in Berlin – the same university where Wolff had studied in the previous century. Müller's research was extremely broad, taking in anatomy, physiology, pathology and embryology. Like Wolff, he studied chick embryos, and although the Mullerian ducts had been seen before, their significance hadn't been understood. Müller saw how these ducts had different fates in male and female embryos: in male chicks, the Wolffian duct persisted to become the vas deferens, and the Mullerian ducts regressed; in female chicks, the Wolffian duct as good as disappeared, while the Mullerian ducts stuck around to become the oviducts.

There was a long hiatus between the discovery of the Wolffian and Mullerian ducts and an understanding of how their different fates were controlled. These pieces of the puzzle would only fall into place after the genetic revelations of the twentieth century: the discovery of the sex chromosomes, the elucidation of the structure of DNA, and the ability to decode individual genes.

There's a gene on the Y chromosome snappily known as SRY – which stands for the 'sex-determining region of the Y chromosome'. It works (along with other genes, because nothing is simple in the genetic conversation that takes place during embryonic development) to prompt differentiation of the testes. Inside each testis, certain cells begin to produce a substance known as anti-Mullerian hormone, or AMH for short. This hormone spells the end for the Mullerian ducts in male embryos. Other cells in the testis begin to produce testosterone, which virilises the Wolffian ducts, urging them to develop into the vas deferens. Testosterone also masculinises the developing external genitals.

An ovary-determining gene has been found. The protein product of this gene inhibits 'male' genes and leads to a feminisation of genitals, but exactly how it does this is still being worked out. Further downstream in the process, hormones produced by the ovary encourage the development of the other female reproductive organs and feminise the external genitals.

Chick embryos proved very useful for finding out general principles about the origin of reproductive organs, but later development of the female reproductive tract in birds is idiosyncratic. In most other vertebrates, from sharks to humans, both Mullerian ducts stick around, usually forming two separate oviducts which open into the cloaca. In birds and crocodiles, the embryo starts off with a pair of Mullerian ducts, but only one differentiates into an oviduct, while the other degenerates.

The reproductive tract of a hen must provide a place for the sperm to meet the egg, as well as providing extra coverings for the fertilised (or even unfertilised) egg. As the hen's egg (which confusingly looks just like the yolk of a hen's egg) rolls down the oviduct, it gets covered in albumen (egg white) and then in the hard shell.

In placental mammals, the female reproductive tract has to do much more. The fertilised egg is not going to be laid, it's going to be retained inside the female body until the baby is ready to be born. The reproductive tract must include somewhere for the embryo to develop into a full-term fetus. The Mullerian ducts stay separate in their uppermost portions, forming a pair of oviducts (or Fallopian tubes, after the sixteenth-century

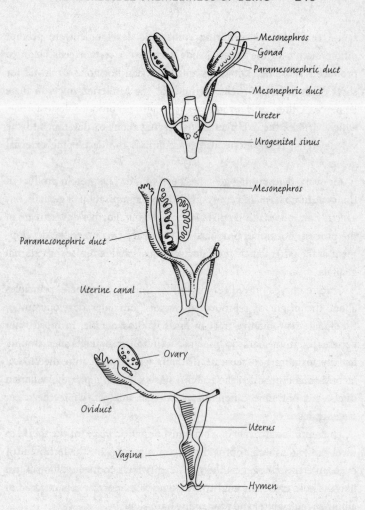

A female human embryo also starts with a unisex kit of parts (top diagram), but this time it's the paramesonephric (Mullerian) duct which persists and develops into the oviducts, uterus and upper vagina (lower diagrams)

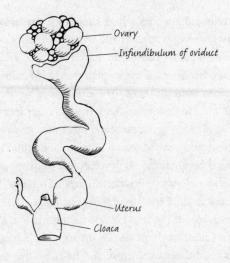

The single ovary and oviduct of a hen

Italian anatomist who first described them). But lower down, they fuse together to form a single vagina, and between that and the oviducts, a uterus. Female marsupials like kangaroos and opossums have paired vaginas (and the males have matching, forked penises) and two completely separate uteri. Most placental mammals have a 'two-horned' or bicornuate uterus (and usually give birth to at least two babies at a time, if not whole litters of them), but there are a few groups where the Mullerian ducts fuse completely in this region, to make a single uterus with no horns. Those groups include some bats, monkeys, apes and humans – all animals which tend to be pregnant with just one baby at a time.

The uterus of placental mammals really is an incredible organ. In humans, this avocado-sized organ stays that way until a new embryo implants inside it – then it grows and grows. At full term, the human uterus swells until it reaches right up to the bottom of the sternum or breastbone. The uterus doesn't simply stretch to this size, it *grows*, piling on muscle which it will need to push the fetus out at the appointed time.

Inside the uterus, although it's a long time since our female ancestors laid their unfertilised eggs straight into water, the fetus is developing in its own primordial pond of amniotic fluid. The invention of the amniotic sac was another essential step which freed land animals from the need to return to water to breed. The embryos of reptiles, birds and mammals still need to develop in fluid. In egg-laying animals the fluid is contained inside the egg; in placental mammals we keep that pond inside the female body, in the womb. The amniotic fluid in which the embryo swims is important to its development in many ways. It cushions the developing embryo inside the mother's body. The embryo 'breathes' amniotic fluid into its developing lungs, swallows it, and absorbs it into its blood. The fetal kidneys filter the blood, making urine (which contains none of the toxins in the urine of babies or adults – those substances are removed from the fetus by the placenta) and passing that urine replenishes the amniotic fluid. If the embryo's kidneys aren't working properly, the volume of amniotic fluid is depleted, the human embryo struggles to grow its lungs, and its restricted limbs grow stunted and twisted.

MIND THE BOLLOCKS: THE INCREDIBLE MIGRATING TESTIS (AND OVARY)

When you were an embryo, your gonads appeared high up on the back wall of your abdomen, but a moment of reflection will confirm that this is *not* where you keep your gonads now, whatever sex you are. While the kidneys ascend from their starting place in the pelvis, rising up onto the posterior abdominal wall, the gonads do the opposite, sinking down until they end up in the pelvis. If you're female, your ovaries stopped there. If you're male, your testes still had some way to go. Under the influence of various male hormones, including AMH, the ligament suspending the testis from above dissolves away, while a new ligament develops below the testis and precedes its descent. This ligament is called the gubernaculum, a Latin word meaning the 'navigator' (and which also gives us the word 'governor'). The testes start to migrate during the second month of gestation; by the time of birth, the testes of most baby boys will already have reached the scrotum. In a few baby boys, the testes still have some way to go; they usually get there within the next few months, but in a small number, the testes fail to descend.

Why do testes descend in the first place? Compared with women, who sensibly keep their gonads on the inside, men incur a peculiar vulnerability by wearing theirs on the outside. It seems a foolish thing to do. These organs started off life safely concealed in the abdomen, so *why* drag them out into the scrotum where they are so vulnerable to injury? Riding a bicycle, leaping over hurdles, climbing over a farm gate – suddenly these seemingly innocuous activities become laden with risk for half the human population.

The answer seems to be that sperm production works best at a few degrees below body temperature. Warmer temperatures, it is hypothesised, would lead to too much DNA damage and defective sperm. The scrotum, the bag in which the testes dangle, is a uniquely mammalian invention, and it probably evolved along with the development of the ability to maintain a warm body temperature, of around 36–38 degrees Celsius. (Having said that, birds present something of a conundrum as they often have even higher body temperatures and don't seem to have a problem with keeping their testes inside the abdomen.) There are a few mammals who keep their testes on the inside, including elephants and whales, which have a specialised blood supply to cool their gonads, and some insectivores, which tend to have a low body temperature anyway. But for most mammals, including humans, of course, the scrotum forms a handy coolbag for the male gonads.

A failure of the testes to descend out of the abdomen into the scrotum, in the condition known as cryptorchidism (literally: hidden testes) has serious consequences. Sperm production is likely to be adversely affected, leading to infertility. There also seems to be an increased risk of testicular cancer. Even with testes which have successfully made their descent down into the scrotum, there may be other factors which lead to an unhelpfully warm temperature around the gonads. Tight pants, hot baths and saunas, and even sitting down for prolonged periods, can raise the temperature of the scrotum and lead to reduced fertility. It's even been suggested that wrapping up the testes and keeping them warm in infancy can lead to long-term increases in scrotal temperature of a couple of degrees, prompting some to suggest that this could affect the testicular stem cells which produce sperm. But whether nappy wearing in infants really does affect reproductive potential in the long term remains unknown.

In many male mammals, testes are kept squirrelled away until it's the mating season, but in primates testes are kept knocking around inside a scrotum all year round. Sumo wrestlers are purported to be an exception, as they are renowned for keeping their gonads safely out of the way during a fight, by massaging their balls up into their groins in preparation. But I must admit, I haven't heard this directly from a sumo wrestler and it's eminently possible that it's an urban myth. It does appear in print, but in a James Bond novel, and Ian Fleming may not be an entirely reliable source when it comes to biological science.

There's a large amount of variation in the size of testes among different primates, even just among humans and our closest ape cousins. An average pair of human testes weighs around 40g, while gorilla testes are smaller, at about 30g per pair, and chimpanzees are the most well endowed, at 135g. Those differences become even more profound when you account for body size: a male chimpanzee weighs about 40kg, compared with around 70kg for a man and a chunky 170kg for a male gorilla. So chimpanzees are the smallest species, but possess the largest testes. Conversely, a huge, male silverback gorilla has a very modest set of gonads indeed. The size of the testes relates to the amount of sperm produced – in promiscuous species like chimpanzees with multi-male, multi-female mating strategies, competition between males is high. Males compete for access to females, but the competition doesn't stop there: a female chimpanzee might end up with sperm from several mates in her reproductive tract.

So the sperm are competing too, and it's largely a numbers game: the male chimpanzee who has produced the most sperm has the best chance of it being one of his sperm which fertilises the egg. In polygynous gorillas, the competition is fought well before copulation happens. A single dominant silverback fights for the right to a harem of females. He doesn't need to produce masses of sperm; relative to body size, gorilla testes produce a twentieth of the amount of sperm produced by chimpanzee testes. Human testes are middling in size: larger than a gorilla's, but about a tenth the size of chimpanzee testes, relative to body size. This indicates a moderate level of sperm competition in us humans – not as intense as in promiscuous chimpanzees, but more pronounced than in gorillas with their silverbacks and harems. The size of human testes reflects a mating system which is largely monogamous, with a little promiscuity thrown in.

PENISES, CLITORISES AND ORGASM

There's also a huge amount of variation in mammalian penises. Having invented such a wonderful organ, mammals sport penises in a great variety of sizes and shapes, and even with various accoutrements. Some are bifid (matching a double vagina), some long and slender, others short and stout, some have barbs and many have a bone (the os penis or baculum) inside them. The variation is quite astounding, even just among primates.

Clitorises also vary quite a bit among mammals, too. The clitoris of female hyaenas and moles is long and penis-like, and there are elongate clitorises among primates too, in some lemurs, spider monkeys and bonobos or pygmy chimpanzees. The females of many primates even possess an os clitoridis – a bone inside the clitoris which is equivalent to the os penis found in many male mammals.

Many male mammals have penises which are kept tucked away neatly in a furry sheath until needed; in contrast, all male primates (including male humans, of course) have a dangly penis. Whereas most male primates possess an os penis, it's lacking in spider monkeys, woolly monkeys and in men (that is to say, in the vast majority of men – in the medical literature, there are 45 reported cases of an os penis in a man). For human males (as well as those monkeys), achieving and sustaining an erection depends solely on local blood pressure, which is how Viagra works. When the penis is flaccid, the smooth muscle in the spongy tissue inside it, as well as the smooth muscle in the walls of the arteries supplying it, is contracted. Cold weather can lead to an even greater contraction of this smooth muscle, causing the penis to shrink further. During sexual stimulation, signals are sent via parasympathetic nerves in the pelvis, causing the smooth muscle in the arteries to relax, and so the vessels widen, letting more blood flow into the spaces inside the spongy tissue, where the smooth muscle has also relaxed. As the pressure increases, the thin-walled veins – which are the route by which blood leaves the penis – are pressed closed. The blood now has nowhere to go, but it keeps on coming, so the penis swells up. As the pressure inside the spongy tissue rises to about 100mm Hg (that's almost the same pressure as the blood leaving the heart at the height of a contraction), the penis becomes erect. Then the ischiocavernosus muscles overlying the crura of the penis contract, pushing the pressure inside it even higher, making the penis rigid.

As erection depends on arterial blood pressure, problems with the arteries can lead to erectile dysfunction. If the arteries supplying the penis don't dilate enough, there won't be enough blood entering the organ to achieve an erection. The nerves supplying those arteries set off a chain of chemical reactions, producing a compound called cGMP (cyclic guanosine monophosphate, if you're interested), which makes the smooth muscle relax. In the penis, cGMP is broken down by a specific enzyme, and it's this enzyme which is deactivated by Viagra – achieving the almost magical effect of increasing blood pressure in the penis, and nowhere else. Or at least, almost nowhere else. A similar enzyme acts in the retina, so taking Viagra can affect colour vision, and sometimes it can affect the brain and heart too, so like all drugs, it needs to be taken with caution.

Given that the clitoris develops from a homologous set of tissues in the embryo, and it's supplied by equivalent nerves to the penis, it shouldn't come as a surprise that the neural and chemical mechanism of clitoral erection is exactly the same. But it took rather a long time for that to be accepted. The discovery that it was the same mechanism, involving cGMP and similar enzymes to those in penile erection, acting in clitoral erection came just ten years ago. It should have been obvious from the shared embryological development and the shared nerve supply in men and women, but the received wisdom was that sex was somehow more cerebral and mysterious in women and more mechanical in men. Here's a quote from that paper (and again, remember this is only ten years ago):

> Sexual function in females has not been widely studied and is often assumed to be different from that in males, with dysfunction resulting more from impaired desire than from impaired erectile dysfunction.

Elucidating the mechanism of clitoral erection doesn't mean that all cases of sexual dysfunction in women can be attributed to erectile dysfunction, but it suggests that some are and could be treated with drugs targeting the chemical pathway that leads to arterial dilation, just as in men. And it's not just the body of the clitoris which becomes engorged with blood and tumescent, the spongy tissue underlying the labia minora also swells up.

Of course, neither penile and clitoral erection are the be-all and end-all of normal sexual function. There's a lot more to coitus than simply getting the penis and clitoris full of blood. In a man, sperm need to leave the testes and then get mixed with secretions from accessory glands such as the prostate and seminal vesicle in the urethra. This process is known as emission, and just like the change of blood flow in the penis, transforming it from flaccid to rigid, it's governed by the visceral or autonomic part of the nervous system. So too is the final phase – ejaculation – where the sperm-filled semen is pumped along the urethra, all the way to its opening on the glans of the penis and out of the body, in rhythmic waves. In a woman, the erectile tissue forming the clitoris swells and becomes tumescent; the wall of the vagina, which is richly supplied with blood vessels, also becomes engorged, and fluid seeps out from capillaries under the lining of the vagina, lubricating it. Various glands, including some in the neck or cervix of the uterus, add their secretions to this lubricating layer of fluid. At orgasm, the muscles of the pelvic floor and the vagina rhythmically contract.

In both sexes, then, the moment of orgasm is associated with waves of muscle contraction. And it seems that orgasm occurs in response not only to impulses travelling in nerves, but to a particular chemical message circulating in the bloodstream. That chemical message is oxytocin, the 'love hormone'.

Oxytocin is a prodigious hormone. Compared with many messenger compounds in our bodies, it's tiny: a string of just nine amino acids. A very short message then – more of a tweet than a memorandum, perhaps – but a very powerful one indeed. The precise ways in which it works are still largely mysterious, but it's linked to all manner of reproductive functions. It promotes pair-bonding (apparently in humans as well as much-studied prairie dogs), bonding between parent and child, and maternal behaviour. It's also the hormone that drives contractions of the uterus during childbirth and the ejection of milk during breast-feeding. It's both difficult and foolish to link a particular emotion or state of mind to one chemical in the brain, but oxytocin has been implicated in feelings of love, happiness and trust.

Oxytocin is also associated with the physical and psychological experience of orgasm. As sexual arousal soars to a climax, there's a surge of oxytocin, released by the pituitary gland under the brain, which probably helps to cause the muscle contractions down in the pelvis. Oxytocin is also released from the hypothalamus, in the base of the brain, and it's this source of the hormone which probably creates the powerful feeling of happiness and wellbeing associated with orgasm.

Sexual arousal and orgasm are still hotly debated topics, especially when it comes to the fairer, more mysterious, and certainly less well-studied sex. Orgasm in women is still something of a political and biological hot potato.

Two centuries of debate about the female orgasm – whether it originates from stimulation of the clitoris or the vagina – have failed to settle the debate. In the nineteenth century, some male biologists considered the clitoris and female orgasm to be of no utility, as neither appeared to be necessary for conception to occur. Freud considered clitoral orgasms as immature; women became sexually mature when they started having vaginal orgasms instead. In the 1960s, the pioneering sexual scientists William Masters and Virginia Johnson concluded that there was only one type of female orgasm, and it started in the clitoris and spread to the vagina. But in the 1950s a German anatomist had described a richly innervated area in the front wall of the vagina – the 'G-spot' – and suggested it was potentially a trigger for vaginal orgasm. More recent anatomical studies have failed to find anything special about this area of the vagina and the literature on the G-spot is rather full of small studies and anecdotal evidence. However, it's possible that vaginal stimulation works because it leads to clitoral stimulation; by pushing and pulling on tissues which attach to the clitoris, including the engorged crura and bulbs. So instead of the dichotomy between clitoral and vaginal orgasm, current understanding suggests it's all part of the same thing. The anterior vagina (with the urethra embedded in it) and the clitoris and its crura all work together, as a cluster of erectile tissue, and stimulation spills over from one area to another. Unsurprisingly, for something so essential to reproduction, other animals have orgasms too, and although there's been much less research into other species, there's no reason to think that the human experience is unique.

GETTING IT ON

As far as using these bits of anatomy is concerned, in some ways we're quite reserved compared with other primates, and in other ways we're even more profligate. Strepsirrhine primates like lemurs generally have discrete seasons of reproductive activity, but some monkeys and all apes (including us) get it on all year round. In monkeys and apes, reproductive cycles last around a month, including ovulation and menstruation. In many monkeys and in chimpanzees, ovulation is advertised by a swelling and reddening of the perineum, and female behaviour changes as well. In humans, there's little to see by way of advertisement. The perineum doesn't swell up, and even if it did it would be hidden between the legs, as we don't tend to walk around on all fours like our closest ape cousins. But it seems there may be some subtle advertising going on after all: studies suggest that women dress more provocatively around the time of ovulation, and one small study showed that lap dancers' earnings peaked at the mid-point of the monthly cycle (although you could look at this the other way round: they earned less while menstruating).

Compared with most other mammals, including dogs, who are restricted to copulating 'doggy style', humans can do it in an astonishing variety of positions. This is down to our ape ancestry. As brachiators who hang around in trees, apes have a large range of motion in their limb joints, making them free to experiment with a range of different positions, although among non-human apes, the male still often mounts the female from behind. However, orangutans, gorillas and particularly bonobos on occasion will copulate in the 'missionary' position, face to face. It has been noted that the frontal position of the bonobo clitoris and vulva may be an adaptation to this position, as perhaps it might be in female humans too.

Although some religions might baulk at the idea, sex isn't all about procreation – in us or in other animals. Same-sex sexual behaviour is a feature of species right across the animal kingdom, particularly in social species. In bonobos, sex seems to be something that is done in the way of a friendly greeting. You don't need to watch bonobos for very long before something interesting happens – they're at it all the time. Bonobos also seem to use sex to diffuse tension and avoid conflict. A lot of bonobos' sexual activity doesn't involve coitus; there's plenty of rubbing of genitals against each other (in heterosexual and homosexual pairings), as well as occasional oral sex and tongue-kissing.

The elaborate machinery that has evolved to allow a sperm to fertilise an egg in a land animal has turned out to be useful for so much more, including cementing social bonds and also pure pleasure. But while the reproductive organs possess all that extra, wonderful potential, there's still one reproductive function we've yet to consider. The anatomy of the male placental mammal is designed to achieve fertilisation – and there's an end to it. (Although men have a significant role to play in supporting pregnant partners and rearing children, there are no anatomical adaptations for this.) But in females, as we've already touched on, the reproductive system must also provide a home for the developing fetus. But of course, it doesn't stop there. There also has to be a way out for the fetus, because at some point that fetus is going to make an appearance outside its mother's body.

A TIGHT SQUEEZE

It was thought that the timing of human birth was uniquely constrained by the anatomy required for large brains and efficient walking on two legs, but recent research suggests that, once again, we've been too quick and too keen to assume that humans are different from any other animal.

The pelvis is the part of the body where differences between males and females are most apparent, and this difference is even there in the underlying bones, especially in humans. The human female pelvis is much broader than the male pelvis; as well as being the link between the legs and the spine, a site of attachment for hip muscles and for the pelvic floor and all the external genitalia, the female pelvis also has a unique role to play: forming the birth canal.

Giving birth to my first child underlined what I already knew: that the fit between the neonatal head and the maternal pelvis was tight. Giving birth to a second baby confirmed my previous findings. Prior to having my first baby, I went along to antenatal classes, where we expectant mothers were lightly patronised (or should that be matronised) by a midwife brandishing a pair of salad servers and a sink plunger in order to illustrate the concept of forceps and ventouse-assisted delivery. I'm sure some people might have gone away thinking their child might be delivered by a plumber or with a spoon. The same midwife also attempted to demonstrate the 270-degree rotation of the fetal head as it passes through the birth canal. As she shoved the baby doll's head through the plastic female pelvis, she said: 'The mother's pelvic outlet is about 10cm wide and the baby's head is about 10cm too. It just fits through perfectly. Isn't Nature wonderful?' My husband – and I was immensely proud of him at this point – piped up: 'I think Nature would be more wonderful if the baby's head was 8cm across.' Absolutely, I thought. Point well made. The midwife moved swiftly on.

But I knew, way before I got to the delivery suite, just how difficult human birth was set up to be. The birth canal of humans is, as the midwife correctly opined, barely any wider than the baby's head. Under the influence of hormones like progesterone and the aptly named relaxin, various joints in the mother's body loosen up during pregnancy. The pubic symphysis, the cartilaginous joint between the two pelvic bones, at the front, softens up and can stretch a little during labour. The baby's head can also squash a bit. The thin plates of bone forming the vault of a baby's skull are separated by fibrous membranes, allowing the bones to slightly overlap each other as the baby makes its way out into the world, and so as a consequence, many newborn babies have slightly odd-shaped heads, sometimes appearing lopsided, but this usually corrects itself within a couple of days after birth.

As well as the difficulty of human childbirth, there's something else about human babies which needs explaining, and that's the fact that they're incredibly helpless. Most primates are highly developed at birth and able to look after themselves to a large degree. Human babies are, in comparison, pretty useless, which must be due to how much brain growth still has to happen after birth. Whereas chimpanzees are born with brains around 40 per cent of adult size, a human newborn's brain is only 30 per cent of adult size. Human babies can't move around on their own, and they can't hang on to their mothers, they need to be carried around and generally looked after, and it's usually about a year before they master the art of standing up and walking on two legs. The helplessness of human infants has important implications for human society – it could even be the reason why humans have evolved pair-bonding, and in most societies, people other than the parents, including grandparents, also contribute to child-rearing. Although being helpless might seem like a disadvantage, it may be that this paved the way for something which has become an important part of being human. That helplessness guarantees intensive contact between the parents (and other carers) and the baby. The human baby's brain has plenty of growing and developing still to do outside the womb, while surrounded by other humans and immersed in their society and culture. As evolutionary psychologist Michael Tomasello puts it, human babies are born expecting culture just as a fish is born expecting water. In fact, some researchers have suggested that the relative immaturity of the human newborn brain is something which has been *selected for* in evolution. In order to survive, we humans depend heavily on knowledge, behaviour and skills which we learn from each other. We've evolved to be that way: having the capacity for social learning is obviously something which proved advantageous to our ancestors, and in fact it underpins our success as a species. In addition, our brains are exposed to learning from a relatively early stage of development, indeed it's been argued that the need for social learning from a very young age could have been the driving force for human babies being born so early. But there does seem to be a more likely, physical reason for the helplessness of our infants.

The obstetric dilemma (OD) is a very neat hypothesis which purports to explain both why human childbirth is so difficult and why human babies are born so early, and so helpless. It suggests that the female pelvis is in the middle of an evolutionary tug-of-war, being pushed and pulled in different directions by the demands of bipedalism on the one hand, and those of birthing a big-brained baby, on the other. The idea is that the pelvis is limited in width by the functional constraints of walking on two legs; if it got any wider, women wouldn't be able to walk efficiently. But it's also a feature of our species that we have big brains and therefore big-headed babies. So the constrained width of the pelvis has a knock-on effect for childbirth: a human baby must be born before its head grows too large for the birth canal. The OD hypothesis suggests that the female pelvis is uniquely constrained by these antagonistic demands, and that the result is a design compromise. The width of the pelvis meant that a human baby had to be born 'early', while it still, just, fitted

The challenge of human childbirth: fitting a baby's large skull through the mother's pelvis

through the birth canal. One of the resulting conclusions of this hypothesis is that the female pelvis is necessarily less than perfect when it comes to bipedalism – and conversely, that the male pelvis should be a more efficient design. After all, we all know that men can (on average) run and walk faster than women, and with their slim hips, they're presumably more energetically efficient at it as well.

However, in 2012, a paper was published in the *Proceedings of the National Academy of Sciences* which called into question the neat explanation provided by the OD hypothesis, and it suggested another reason for the timing of human birth. It's always exciting when something like this paper is published. You read it knowing that your world view will have shifted a little by the time you're finished. And that's what I love about science, every now and then something comes along which challenges an existing paradigm, turning ideas on their heads, shaking up your thinking.

I travelled to New York to meet the lead author on that *PNAS* paper, Holly Dunsworth, and find out why she thought humans weren't as unusual as we'd all previously thought. Holly told me that she'd started off with some suspicions about the assumptions underlying the OD hypothesis. She was studying the length of gestation in other primates and she found that human pregnancy didn't actually seem to be truncated. Human gestation lasts 38 weeks, which is longer than in any great ape (ranging from chimpanzees, at 32 weeks, to gorillas at 36 weeks). Primates of different body sizes have different lengths of gestation, and working out gestation length relative to body size, humans still come out as having the longest pregnancy. It's actually over a month longer than you'd expect for a primate of our body size. This suggests that gestation *has* lengthened over the course of human evolution. In other words, human babies – compared with other primates – are actually born late, not early. Despite this relatively long gestation, our newborn babies' brains still have a lot of growing left to do, simply because our adult brains are so large. This might leave you wondering why gestation couldn't be pushed a bit longer, so that the brain could have a chance to grow a bit more before the baby popped out. So maybe that's when the OD does come into play, perhaps gestation couldn't be pushed any further because that would mean female locomotion would become hopelessly inefficient.

But just how much less efficient were female pelves than those of their male counterparts? Holly introduced me to a couple of her colleagues who were interested in this question: Herman Pontzer and Anna Warrener. We met in a lab on the corner of a block in Hunter College with panoramic views out over New York. Herman and Anna had roped in a couple of students to run on a treadmill while wearing a mask, in order to enable them to measure their oxygen consumption. The results from that day in the lab nicely illustrated the findings of other studies looking at the efficiency of walking and running: women – even with their relatively wider hips – turned out to be just as economical as men.

I found this really surprising, as I expected the muscles around a broad pelvis to have to work harder. When you walk or run, abductor muscles on the outer side of your buttock, above the leg which is in contact with the ground, have to pull tight to stop the pelvis dipping down on the other side. It's really important to be able to keep your pelvis horizontal like this, as it means that you can easily swing the other leg through to take the next step. Without the abductor muscles working on the hip of the stance leg, the swinging leg would drag along the floor. Weakness of those important abductor muscles, which can occur in polio, for instance, creates a strange way of walking, where the pelvis does tip down as a leg is moved forward. This is called a Trendelenburg gait, after the German surgeon who first described it.

But the studies looking at the efficiency of walking and running in men and women suggest that it's not so simple. Even though those abductor muscles should be having to work harder in a woman with a wide pelvis, compared with a man, there are other things going on which mean that walking and running doesn't end up being energetically more costly in women. Subtle adjustments in the position of the leg during walking might be compensating. There's another reason to doubt that the demands of locomotion, of efficient walking and running, have constrained the female pelvis in particular, in that there is an awful lot of variation in the shape and size of female pelves today. You wouldn't expect quite so much variation if natural selection had been acting strongly to keep things in check.

This discovery – that women were just as efficient as men in walking and running – effectively removes one of the supposed design constraints on the pelvis: it doesn't look like there is any particular advantage in having narrow hips. In addition to that, fossil evidence shows that the pelvis has undergone major changes during human evolution. Compared with older, smaller-brained ancestors, the pelvis of early members of our own genus, *Homo*, shows an absolute increase in width. So the pelves of our ancestors *were* growing wider as brain size increased. All right then, but why haven't female hips got even wider, making birth easier and allowing the baby to stay inside longer?

Before I went to New York to meet Holly and her colleagues, I had spent two weeks collecting and freezing small samples of urine after drinking some isotopically labelled water. I'd posted the samples off to the States, ahead of my trip. Herman analysed my results using the levels of the different isotopes of oxygen and hydrogen in my urine to work out my metabolic rate, and when I arrived in New York, he showed me my data points superimposed on a graph. My metabolic rate was about double what you'd expect – for a non-pregnant woman. But I was five months pregnant at the time: I was providing energy for my developing baby too. Herman told me that when my metabolic rate rose to 2.1 times the normal rate, I would go into labour – my body would no longer be able to provide the energy being demanded by my developing baby.

This prediction was based on good data. The precise mechanism for timing birth in humans (or any other mammal) remains elusive; indeed, it's been called 'the greatest unresolved question in reproductive biology'. But, although we might not know the details of this mechanism – exactly which chemical messages are involved, and how – it seems that the moment may be determined by this energy balance. Birth occurs when the energy demands of the fetus threaten to exceed maternal capacity for supply. A mother's metabolic rate does exactly what mine had done:

by about halfway through the pregnancy, it rises to twice the normal, basal metabolic rate (BMR), and by nine months, the energy demands of the fetus push it even higher, close to 2.1xBMR. The crossing of this energy rubicon coincides with the end of gestation.

So it looks as though it could be this 'energetic crisis' – not pelvic size – which limits the length of pregnancy. There are some potential problems with the hypothesis; the baby, of course, continues growing and demanding energy once it is born, and in mammals, including us, that energy is still provided by the mother in the form of milk. In fact, a growing baby demands *more* energy than it did as a fetus because its metabolic requirement just rises and rises. But even though the mother has to provide more energy to her baby once he's born, the *rate* of increase in the mother's metabolic rate slows. Perhaps it's the steeply increasing energy demand of the fetus in the womb that causes the 'energetic crisis' – the mother's body can't respond fast enough. After birth, the baby's energy demand will keep on increasing, but at a slower rate that the mother can keep up with.

This 'energetic crisis' hypothesis is appealing because it offers an explanation of the timing of human birth which isn't necessarily unique, and the same mechanism could be acting in other placental mammals, too. Across mammals, gestation length and the size of newborn infants is related to the mother's body size.

This new 'energetic crisis' hypothesis seems to explain the timing of human birth better than the old obstetric dilemma or OD hypothesis. It means that human babies are born at a point where the energy demands of the fetus are rising too fast. As shown by the fossils, the hominin pelvis became progressively wider to accommodate bigger-brained babies, but only up to a point. Even if the pelvis got wider still, babies wouldn't stay inside any longer because it's that energy crunch which determines when they're born. But there's still something it can't explain, and that's the difficulty of human birth.

Firstly, are we right in thinking that birth is much more difficult in humans compared with other species? Well, yes, compared with most mammals, but if we take a look at primates, then quite a few of them have a challenging time of it, too. It's difficult to draw these comparisons, as relatively few non-human primate births have ever been witnessed. But, from what has been observed, it seems that chimpanzees, gorillas and orangutans, with their wide pelves, have relatively easy births. Monkeys, however, are worse off. Quadrupeds tend to have a narrow and deep chest, and the pelvis follows that shape, which means that quadrupedal monkeys have a narrow pelvis, making for difficult childbirth. In marmosets and squirrel monkeys in particular, the mother's pelvis is only slightly larger than the fetal head.

So it's not that all mammals have it easy while humans, uniquely, have a difficult time; it seems instead that there's a spectrum of difficulty. Having said that, it still seems that human childbirth is certainly out at one end of that spectrum – it's longer and apparently more painful than it is in any other primate. Despite a tight fit between the fetal head and the pelvic outlet, monkeys are perhaps a little better off as they have narrow shoulders. Human newborns, with relatively broad shoulders as well as large heads, must pass through three rotations, spinning through 270 degrees on their way through the birth canal. Because of these rotations, human babies are usually born face down, whereas monkey neonates are more usually born face up. A monkey mother can help her baby out into the word by pulling up on its emerging head, but pulling upwards on the head of a human infant as it comes out – flexing its neck backwards – could be very dangerous. Unlike other primates which seek seclusion during childbirth, human mothers tend to seek assistance. Midwifery, then, is probably very ancient indeed.

As well as being painful, childbirth can be risky. A mismatch between the size of the mother's pelvis and the fetal head can result in obstructed labour, where the baby *cannot* be born naturally. Obstructed labour places the lives of both mother and baby at risk. Despite the fact that human mothers seek assistance during childbirth in most human societies, obstructed labour is still a significant problem in many human populations.

Across the world, obstructed labour, or dystocia, is estimated to affect around 3–6 in every 100 births, but a recent study in West Africa found a much higher incidence, with obstructed labour in one in five births. As dystocia presents such a such a threat to survival – of both mothers and babies – it seems strange that female pelves haven't evolved to be wider.

Having dismissed the OD as the reason for the timing of human birth and the helplessness of human babies, perhaps that hypothesis still has something useful to say about the difficulty of childbirth. Locomotion may currently be as efficient in women and men, but perhaps if the female pelvis were to get any wider, efficiency *would* end up being compromised. Yet this doesn't seem to ring true either. Look at the effects being weighed against each other: on the one hand, a wider pelvis could mean that walking and running require a little more energy; on the other hand, a more energetically efficient, narrower pelvis would place the lives of mothers and babies at risk. It seems an unlikely pay-off. The consequences of a pelvis too narrow to birth a baby far outweigh any locomotor disadvantage. Natural selection is very unlikely to have produced a constraint on pelvic width which would so effectively compromise survival.

So how can we explain such high rates of dystocia as we see in some populations today? Even that average incidence of 3–6 per 100 births sounds high. It's possible that natural selection has just not had enough time to act yet: perhaps difficulty in childbirth is a very recent phenomenon, and not something which our ancestors would have experienced anywhere near as often as it's seen today. There seems to be some support for this idea, from both archaeological and current data.

Although there's very little in the way of robust data, anecdotal evidence from contemporary hunter-gatherer communities suggests that dystocia is very rare. It's also rare in the archaeological record: a handful of examples of female skeletons, from different sites, through the ages, have fetuses within the pelvis. The picture is very different in contemporary agricultural communities, particularly where childhood malnutrition affects the growth of girls, meaning that women tend to be shorter, with smaller pelves. In these same communities, women's access to obstetric care tends to be limited.

Dystocia is much less common for women giving birth in today's affluent societies. Problems might arise because of mothers consuming a diet with a high glycaemic index – in other words, a diet that tends to lead to high levels of blood sugar. That, in turn, tends to push up the birth weight of babies, so the problem here could be less about pelves which are too small, and more about babies that are too big.

In the poorer countries where malnutrition is leading to high rates of dystocia, and where there's limited access to obstetric care, it's possible that natural selection may be acting right now, favouring women who have wider pelves despite malnutrition, or smaller babies. In richer countries where obstetricians are waiting in the wings when birth becomes difficult, natural selection is side-stepped. Mothers and babies still have a high chance of survival even when a mismatch between the width of the mother's pelvis and the size of the baby's head makes a vaginal delivery dangerous or even impossible. The scalpel of the obstetrician, and indeed the helping hand of the midwife, is sharper than the scythe of natural selection.

Across human societies today, perhaps we can blame difficult birth on our ancestors. Having help during labour would have lessened the selective pressure against big-headed babies or narrow pelves in mothers. So we have difficult births now *because* we're cooperative: we need help during birth because our ancestors *had* help during birth. Midwifery has been called the oldest profession, and it seems that once midwives were needed, it was guaranteed that they would always be needed.

We've spent a significant amount of time delving into the pelvis, finding evolutionary clues which hark back to fishy and amphibian ancestors, finding out how land animals overcame the challenges of conceiving a baby and discovering that the size of the human female pelvis doesn't seem to limit the amount of time a fetus can stay on the inside. You may even know a little more about the clitoris and the penis than you ever did before. But apart from housing the reproductive organs and forming the birth canal in a woman the pelvis has another job – it forms the attachment between your trunk and your legs.

ON THE NATURE
OF LIMBS

Fins, limbs and ancestors

'Once we were blobs in the sea, and then fishes, and then lizards and rats and then monkeys, and hundreds of things in between. This hand was once a fin, this hand once had claws!'

TERRY PRATCHETT

BUDDING LIMBS AND GROWTH PLATES

To find out where your own limbs come from, we have to go back to an early stage of embryonic development. Just over four weeks after conception, the embryonic you has rolled up, turning from a flat, three-layered disc into a stack of nested cylinders. You can see the thickened discs on each side of your head which will form the lenses of your eyes and the fluid-filled labyrinths of your inner ears. Your neck is flanked by a series of gill-like branchial or pharyngeal arches. A chain of somites is visible on each side of your body at the back, and you have two pairs of bumps which are destined to become your arms and legs. To begin with, these limb buds are simply a core of loose embryonic connective tissue which originated from the middle, mesoderm layer of the trilaminar germ disc (the 'jam' in that embryonic sandwich), covered in an outer coat of ectoderm. By week six, your limb buds have grown longer and their ends have flattened to form hand and foot plates. During week seven, cells begin to die in the hand and foot plates.

This is a normal part of development; sculpting a human body involves processes of cell death as well as growth and proliferation. Cell death is important because it reduces the amount of genetic information needed to construct a body. In fact, it's a fundamental process in embryonic development in any animal with a complex body. In many cases, it's much simpler to generate too much tissue and trim back to what's required, rather than carefully programme a complex pattern. A good example of this can be seen in the developing nervous system: the brain of a newborn baby contains about 90 billion nerve cells or neurons, then branches and indeed entire neurons are heavily pruned back until only about half that number is left in an adult human brain. The pruning process leaves only those neurons and synapses which are proving useful; it's a physical manifestation of learning, but it's also much easier to make far too many neurons to start with and then reduce them to what's needed after they've established their connections, rather than to carefully programme the connections of each. My three-year-old knows this principle. If she's making a picture using glue and glitter, she'll start by drawing patterns with the glue. Next, rather than carefully pouring the glitter along the glue trails, she'll shake out glitter across the whole sheet of paper, then tip off the excess. In the development of the eye, dying cells allow the developing lens vesicle to detach from the overlying surface ectoderm. Programmed cell death – cells committing suicide – carves tubes out of solid rods of tissue, which is precisely how blood vessels first form throughout the embryo. The chambers of the heart are sculpted out by cell death as well built up by cell growth, and in the tips of the developing limbs in humans, mice and birds, cell death allows the digits to separate. If that programmed cell death fails, fingers and toes may remain fused together.

As the external appearance of the limb changes from a bud to a recognisable miniature arm or leg, things are equally dynamic and interesting on the inside, too. Within the mesoderm core, embryonic cells clump together and differentiate into cartilage cells. By week six, the limb

The developing arm and hand in a human embryo

contains tiny cartilage models of the bones it will eventually possess. Cell death kicks in again, this time to form the joints between the cartilage models. Cells surrounding the newly formed joint cavity will become the cartilage lining the surface of the bones at the joint, as well as the membrane which holds in the slippy, synovial fluid which lubricates the joint.

During the eighth week, some of the cartilage in the middle of each of the model 'bones' has started to change to actual bone tissue, and this ossification spreads until, by birth, only the very ends of the 'bones' are left as cartilage. After birth, islands of ossification also appear in these ends, but they don't join up with the bony shaft until many years later. The secret to being able to grow your bones quickly is to keep a little bit of cartilage in them. This cartilage takes the form of a disc, sandwiched between the shaft and the end of the long bones in your limbs. This disc is known as the 'growth plate' and it only disappears when the end of the bone fuses with the shaft, once your bones are fully grown towards the end of your teenage years. Once that cartilage layer is gone, the bone no longer has the capacity to grow in length. In some bones, this fusion occurs very late indeed – the growth plate at the inner end of your clavicle (or collarbone) stays open and cartilaginous well into adulthood, and doesn't fully fuse until you're about 30 years old.

As the growth plates fuse at slightly different times, and in a fairly regimented programme (although everything in biology is subject to variation), it means that checking up on growth-plate closure can be a useful way to investigate if a child is growing normally. It's also extremely useful when it comes to determining the age of juvenile archaeological skeletons. Here's an example from a Viking site on the island of Anglesey in North Wales.

The site is on a farm near the northeast coast of Anglesey, with wonderful views of Snowdonia on the mainland. It was discovered by a metal detectorist who found a Viking coin there, prompting an investigation by archaeologists from the National Museum of Wales, led by Dr Mark Redknap. The site would turn out to be the first evidence of a

Viking settlement in Wales, perfectly placed as a stop-off on trade routes between Scandinavian homelands and settlements in northwest England, west Scotland and Ireland. Excavations were carried out on the site every summer for more than a decade, and I enjoyed wielding a trowel on many of the digs, as the resident osteoarchaeologist. Several burials were discovered over the many seasons of work at the site, including five skeletons in a ditch at the edge of the settlement. Burial is perhaps too formal a term for the way in which the remains of those bodies ended up in that ditch: they were clearly thrown in in a less-than-respectful manner. One adult male skeleton had his hands positioned so close to each other, under his trunk, that it seemed most likely that his wrists had been tied together when he'd been buried. Whoever these people were, they had probably met with an untimely death. Two of the skeletons had belonged to juveniles.

I helped to excavate the skeletons, but my work continued long after we'd backfilled the trenches. Back at the Museum of Wales, in Cardiff, the skeletons were meticulously cleaned and the fragmented bones were pieced back together. Then it was my turn to pore over the bones, to try to find out as much as I could about the people whom these remains represented.

Among the bones of the skeleton named 'burial 2', the ends of many of the long bones were unfused. The blunt ends of the shafts and the separate, cap-like ends of the bones had the typical, rippled, undulating appearance of bone where it sits next to an unfused cartilaginous growth plate. Cartilage is soft and it disappears in the burials, leaving just the bony parts of the skeleton behind. I looked carefully at the pattern of growth-plate fusion. The lower end of the humerus (upper arm bone) was unfused – it has fully fused by around seventeen years of age; the bases of the finger bones were also unfused – these are fused by age sixteen; but the three bones forming the hip socket had fused together, and the earliest they start to fuse is at around eleven years. These factors indicated to me that the skeleton was from someone who was between eleven and sixteen years old. This skeleton also had a well-preserved dentition, and careful inspection of the teeth allowed me to refine this age estimate. The second molars had erupted into the mouth, while the third molars (the wisdom teeth) were unerupted; the crowns of these teeth were complete but the roots had not yet formed. This meant the child was probably between twelve and thirteen years old when they died.

Although I like to think I'm an objective scientist, I'm also human, and it's impossible to look at remains like this – especially when it seemed that the deaths had most likely been violent – and not feel sad. The end of this child's life, along with the other individuals thrown into that ditch, is shrouded in mystery, but the archaeologists are still working on this centuries-old cold case in an attempt to understand what happened in that corner of Wales so many moons ago. Were these local casualties of a Viking raid? Or Viking settlers killed by native Welsh? Or victims of in-fighting? Chemical analysis of tooth enamel might help to solve the question of where these people originated, and it suggests that they may have been Vikings, but the precise circumstances of their deaths is hard to elucidate from these long-dead remains. I have to leave the story there. My job, as an osteoarchaeologist, was over after I had measured the bones, made my notes and photographed the skeleton. I put the bones away into their own acid-free cardboard box and moved on to the next one.

THE MYOTOME DANCE

It's all very well filling up your embryonic limbs with bones, but they're not going to be much use if you don't have any muscles to move them. That is, after all, one of the major reasons why you have bones in the first place – they provide a system of levers for your muscles to work on. The muscles in your limbs migrated there from the segmented chunks of tissue flanking your developing vertebral column when you were a tiny embryo. To begin with, they stay as separate segments of muscle tissue, but then those blocks of muscle start to fuse and split, forming the definitive muscles of the limbs, and most end up containing tissue which originated from two or three (or more) of those original segments or somites. In the front of your adult arm, your biceps brachii muscle (or biceps for short) contains muscle which originally came from the fifth and sixth cervical somites. The pattern of innervation of your adult muscles recalls that embryonic origin; the musculocutaneous nerve which supplies your biceps muscle contains motor nerve fibres which have entered it from both the fifth and sixth cervical spinal nerves, in the neck. As you move lower down, into the hand, you find muscles which are innervated by the eighth cervical (C8) and the first thoracic (T1) spinal nerves. If you flex your fingers, it's mostly C8 which is carrying the nervous impulse down to muscles; when you spread your fingers apart, it's T1. The term 'myotome' refers both to the part of a somite which will become skeletal muscle, and to the group of muscles which, in the adult, share a common ancestry in one embryonic myotome and are therefore innervated by a single spinal nerve.

Medical students learn the pattern of myotomes in the limbs, not so that they know where muscles came from in the embryo, but so that they know which muscles will be affected if a particular spinal nerve is damaged. One of the *aides memoires* to learning these associations between particular spinal nerves and specific muscles and their actions is to learn a 'myotome dance'. The choreography is not all that complex, involving a string of consecutive spinal nerve roots and their associated movements in the upper limb (arm and hand) and the lower limb (which is the name we anatomists persist in giving to the leg). There are plenty of examples of myotome dances on YouTube.

These patterns of muscles and nerves in your adult limbs, which hark back to segmental origins in the embryo, are not exclusive to humans by any means. The broad patterns are something you share not just with other mammals but with all tetrapods. As a member of this group of vertebrate land animals, your limbs are lined up with your spine in a particular way: your upper limbs (equivalent to forelimbs in four-legged animals) sprout from your body level with the transition from the cervical to the thoracic spine. Your lower limbs (equivalent to hind-limbs) are aligned with the change from lumbar spine to sacrum – and that's true whether you're an amphibian, reptile, bird or mammal. It's clear that the 'design' of a tetrapod is constrained in this way – you don't get any tetrapods with limbs halfway down the thorax or sprouting at the base of the tail. This could be a functional constraint: animals with limbs in weird places wouldn't move well and would be weeded out by natural selection, or it could be a constraint that's even more funda-mental, deeply embedded in your genome. As Denis Duboule's research group discovered, the *Hox* master-control genes which control segmen-tation along the body axis also determine the patterning of the limbs. Perhaps, if you mucked around with the *Hox* genes to get limbs in different places, you would upset the whole body plan so badly that there would be no hope of making a viable organism.

As we've now seen with so many structures, bits of anatomy don't just spring into existence. Evolution loves to recycle, to repurpose and to tinker. Tetrapod limbs didn't sprout from bodies at the moment when our fishy ancestors made their transition onto land, those ancestors already possessed appendages in the form of fins.

FINS AND LIMBS

The link between fins and limbs is not a new revelation. The English biologist and comparative anatomist Richard Owen wrote about it in his book, *On the Nature of Limbs*, published in 1849, drawing comparisons between the limbs of mammals, birds and the fins of fish.

Richard Owen was a prominent biologist in the nineteenth century who made important contributions to the field, but he ended up being a controversial figure, apparently being almost universally disliked. Darwin, usually such a mild-mannered Victorian gentlemen, wrote that he hated Owen. Richard Owen was destined to become a doctor, studying medicine in Edinburgh and then London, but after qualifying he turned his back on a clinical career and embarked on anatomical research, becoming professor and conservator in the museum of the Royal College of Surgeons. In 1856 Owen moved on to become the superintendent of the Natural History department of the British Museum, and oversaw the transfer of that department to its new home in South Kensington – now known simply as the Natural History Museum.

Owen's research was extremely wide-ranging. He studied and wrote papers on invertebrates including sponges, horseshoe crabs and the pearly nautilus, as well as all manner of vertebrates. He recognised that some ancient reptiles had mammal-like characteristics, and it was Owen who coined the term 'dinosaur', meaning 'terrible lizard'. You can see Owen's vision of what dinosaurs would have looked like, in the flesh, at Crystal Palace Park, where several of the reconstructions he helped to create for the Great Exhibition are still roaming around and swimming in the lake.

Over the course of his career, Owen seems to have had a knack for making enemies. He passed off the work of other scientists as his own, and was booted out from the Royal Society's Zoological Council, fatally damaging his reputation. He seems to have been frustrated that someone whom he probably saw as a junior in the field should have written such an influential book about evolution, and published an anonymous review of Darwin's *Origin* in the *Edinburgh Review* in 1860. It makes for an uneasy read. In it Owen was keen to point out that Darwin was not the first scientist to discuss the 'transmutation of species', and he suggested that many 'younger naturalists have been seduced into the acceptance of . . . "natural selection".'

Perhaps I'm being overly sensitive on Darwin's behalf, but I can't help but feel there's more than a subtle note of derision in: 'No naturalist has devoted more painstaking attention to the structure of the barnacles than Mr Darwin.' Owen proceeds to pull out what he sees as the real contributions to natural history in Darwin's book: observations on barnacle anatomy, the behaviour of ants and bees, the possibility of wading birds transporting seeds in the mud on their feet, and character-istics of domestic pigeons. 'These', writes Owen in his review, 'are the most important original observations . . . they are, in our estimation, its real gems, – few indeed and far apart . . .' Ouch!

Owen doesn't deny evolution as such in this review, but he certainly questioned the mechanism proposed by Darwin: natural selection. Instead, Owen saw the generation of species as a process which was pre-ordained, governed by 'a continuously operating creative force'. This idea directly opposes Darwin's theory of evolution by natural selection, in which nothing was pre-planned, and changes came about simply because of the way organisms interacted with their environment. Owen was particularly uncomfortable with the idea that large changes could occur in animals: that one species might be able to change into another species.

But eleven years earlier, when he wrote *On the Nature of Limbs*, Owen had seemed more open to the idea of organisms changing considerably over time. At the end of the book he wrote: 'Nature . . . has advanced with slow and stately steps, guided by the archetypal light . . . from the first embodiment of the Vertebrate idea under its old Ichthyic [fishy] vestment, until it became arrayed in the glorious garb of the Human form.' While he had a real problem with the idea that humans had descended from apes – in fact he went so far as to put us humans in our very own subclass of mammals – it seems he could accept the idea that, much more distantly, humans had descended from fish. It's clear from Owen's words that he believed in a *scala naturae*, and that humans repre-sented its apogee. To him, the similarities in the limbs of different animals represented variations on a theme, or variations on an arche-type, that were somehow embedded in a preordained plan. Although Owen wasn't a creationist as such, he believed that evolution had been guided, if not by a deity, then by a force he characterised as 'Nature'.

Although Owen's conclusions now sound outlandish, his observa-tions were careful and he recognised the deep similarity between the limbs of land animals and the fins of fish. He also recognised that he wasn't the first to notice this similarity. He wrote:

> The 'limbs' to which the limits of the present Discourse confine its application, are those of the Vertebrate Series of animals; they are the parts called the 'arms' and 'legs' in Man; the 'fore-' and 'hind-legs' of Beasts; the 'wings' and 'legs' of Bats and Birds; the 'pectoral fins' and 'ventral fins' of Fishes. This special homology has been long discerned and accepted . . .

Owen goes on to describe the underlying similarity between limbs as apparently different as the oar-like 'fin' of a dugong, the forelimb of a mole, the wing of a bat, the legs of a horse, and the limbs of a human. He points out that the outward appearance and the function of these various limbs can be strikingly different, but says 'We cannot be surprised at this; it could not be otherwise; the instrument must be equal to its office.' But then he lists various devices which humans have invented in order to move through water and air, on and under ground: 'the boat and the balloon, Stephenson's locomotive engine and Brunel's tunnelling machinery', pointing out that there is no commonality in design linking those devices. If the *purpose* of animal limbs was the sole driving force in their design, he argues, you'd expect them to be much more diverse and not to have those deep, underlying similarities.

Owen drew attention to the similarity of limbs *within* an organism: 'Every one may see that the thigh answers to the [upper] arm . . ., the [lower] leg to the fore-arm, the ankle to the wrist, the five-toed foot to the five-fingered hand . . .' Inside the limbs, the bones of the arm and the leg are also surprisingly similar.

Making comparisons between species, he points out that the same elements are present in the arm of a human and the forelimb of a horse, albeit with the digits reduced to just one in the horse. The similarities are there, too, in the wing of a bat. Now we understand that the homologies in the limbs of various animals reflect a deep homology within their genetic codes – in the *Hox* genes. Owen saw deep similarities in the fins of fish too, even seeing homology between fingers in a hand and rays in a fin. He was later to be proved wrong in this instance – the hand or foot at the end of a tetrapod limb does seem to be a new invention, involving a completely different pattern of *Hox* gene expression at the tip of the developing limb, rather than a variation on what already existed in the fins of fishy ancestors. It might seem almost impossible for tetrapod ancestors to have evolved new digits on both the fore- and hindlimbs at the same time, but it seems that a single genetic 'switch' operates – via *Hox* genes – to produce both fingers and toes.

So when did this astonishing transformation take place? To answer that, we have to turn to the rocks and look for fossils of ancient animals which were taking steps towards the unexplored land.

A GREENLAND SAGA

In 1987, the palaeontologist Jenny Clack was searching for fossils in Greenland. And she knew what she was looking for – fossils dating to 365 million years ago. In the 1930s, Swedish palaeontologists had found some intriguing fragments of skull in East Greenland, and the new species had been named *Acanthostega* ('spine armour', after the bony spikes sticking out of its skull). In 1971, a geology student had come back from East Greenland with some fossils which also turned out to be from *Acanthostega*. Jenny Clack went back to the site as part of a joint Danish and British team, hoping to find more of the animal. And they did, not just skulls this time but, incredibly, several almost complete, articulated fossilised skeletons of the beast.

The reason that this creature is so important is that it gives us an amazing insight into that vital transition between fish and the first land-living vertebrates. *Acanthostega* has around two-thirds of the 41 defining anatomical characteristics of a tetrapod, but it's also very fishy.

Acanthostega was a somewhat bizarre-looking creature. It looks, to me anyway, like a giant newt. It was about a metre long, with a flattened head. The connection between the head and the neck was an odd one, it looks like it relied on an extension of the notochord into the braincase. That's something which is only seen in some fish and embryos of other vertebrates today; in living tetrapods, including you, the back of the skull articulates with the spine. Fishy features included a skeleton around the pharynx which clearly shows that *Acanthostega* possessed gills. Like fish, and unlike most other tetrapods, the bodies and the neural arches of the vertebrae were very similar all the way along the spine. But despite an array of fish-like features, *Acanthostega* also possessed limbs. Proper limbs, with fingers and toes. Lots of them.

All tetrapods around today have five digits or less at the end of their limbs. So it seems reasonable to assume that we've all descended from a five-fingered or 'pentadactyl' ancestor.

(That's interesting in itself, because if we consider an animal which still had five fingers, we'd call this a 'primitive' characteristic. An animal that possesses a reduced number of digits exhibits a 'derived' condition. You can see where this is going. You might think of yourself as 'highly evolved' but what exactly does that mean? In terms of fingers and toes, at least, you're very primitive indeed. A horse, with a single digit at the end of each limb (and rather appropriately, it is the middle one, the *digitus impudicus*, meaning that horses are walking around giving us the finger) exhibits that derived condition. You and I are inescapably primitive in this regard. I hope I haven't offended you, but I also hope that this little aside may persuade that 'primitive' and 'derived', while useful terms when you're reconstructing a tree of life, are not imbued with any sense of value. A derived characteristic is not, in itself, 'better' than a primitive one.)

So it was a real surprise when palaeontologists found *Acanthostega*'s fingers – all eight of them. A beautifully preserved fossil of an entire, articulated forelimb ended in this embarrassment of digits. The back feet are less well preserved, but it looks as though this early tetrapod had at least eight toes as well.

The forelimb was remarkable in other ways, too, there's no true wrist joint, which makes it impossible for this to have been a weight-bearing limb. This was contrary to expectations, what, after all, were limbs for if not to allow an animal to walk out of the water onto the land? *Acanthostega* forced a rethink as it looked as though the earliest limbs had evolved not for walking on land, but for crawling and paddling around in water. Once again, we're reminded of evolution's propensity towards recycling and repurposing bits of bodies.

Since the discovery of *Acanthostega*, many other early and near-tetrapods have been discovered, so that's now a fairly bushy tree of known species around this key transition from aquatic to land animals.

A few early tetrapod species possess more than five digits per limb, with this variability later settling down to the familiar and more compact five by the time properly weight-bearing limbs develop. But before that commitment to a land-based existence, the earliest tetrapods, like *Acanthostega*, seem to have lived in a variety of environments, from freshwater lakes to brackish lagoons and estuaries. Combining what's known of the anatomy of *Acanthostega* with reconstructions of the environment it lived in helps us to picture how this animal would have lived. Imagine this ancient, newt-like creature, with a top-to-bottom flattened head and eight fingers or toes at the end of each limb. It lives in warm, shallow, tropical swamps and uses its limbs to paddle around, making its way through dense weeds – very much like a huge newt among pondweed. It creeps up on smaller creatures floating near the surface, breaking the surface of the water at the last minute and snapping them up, chewing its meals briefly with two rows of sharp teeth before swallowing them down. It's well adapted to these shallow swamps, but its limbs wouldn't bear its weight if it did try to haul itself out. When they die, some of these creatures sink into the mud that forms the floor of the tropical swamps they inhabit. Around 365 million years later, that mud is rock in East Greenland, and palaeontologists are chipping away at it to release the early tetrapods from their petrified tomb.

Despite the fact that *Acanthostega* is such a distant relative, and an aquatic one at that, there's another important change in its body beyond the transformation of fins into limbs. The evolution of limbs from fins involved reorganisation of the way these appendages attached to the rest of the skeleton. As pectoral fins became forelimbs, they detached themselves from the back of the skull, where they'd been attached in fish. Conversely, as pelvic fins became hindlimbs, they attached themselves very firmly to the spine. It used to be thought that the fishy ancestors of tetrapods had stronger pectoral fins compared with their pelvic fins, and that there was a shift as amphibians evolved, with the hindlimbs becoming more powerful. But the discovery of pelvic bones from the fossil fish *Tiktaalik* casts doubt on this 'front-wheel-drive' hypothesis, or at least, moves the shift earlier, before tetrapods appear. *Tiktaalik*'s pelvis is about the same size as its pectoral girdle (equivalent to your shoulder blade and collarbone), so the pelvic fins were already becoming more powerful before limbs evolved.

Most tetrapods today still make use of that rear-wheel drive. Very generally, the way your limbs attach to your skeleton follows a tetrapod pattern. Your upper limb, equivalent to the forelimb of four-legged animals, is loosely connected to your chest via the pectoral girdle (consisting of the clavicle and scapula). In contrast, your lower limb or leg, equivalent to the hindlimb of quadrupeds, is very firmly attached to your spine, via a pelvic girdle.

Just think about the way your shoulder and your pelvis are connected to the skeleton of your trunk. Your shoulder blade 'floats' on the back of your ribcage, attached to it only by muscles and the narrow strut of the clavicle or collarbone to your sternum or breastbone in the front. Place one hand on the back of the opposite shoulder, where you should be able to easily feel the spine of your scapula. Now move your free arm around – raise it up and down, push it back and forwards – and you should be able to feel your scapula moving around on your back. In contrast, your pelvis is fairly firmly welded to the bottom of the spine – to the sacrum. It's impossible to move your pelvis from side to side without moving your spine too. A quick experiment, wobbling your hips from side to side, with one hand on the small of your back, should confirm this. The sacroiliac joints, between the upper part of the pelvis, called the ilium, and the sides of the sacrum, are large joints, stabilised by chunky liga- ments. Those joints need to be large and stable because they are heavily loaded in standing and walking, and even more so when you run or jump. They're the point of contact between your lower limbs and your spine. They have to withstand large forces because, as REM almost said (they don't seem to have been that hot on anatomy), your legs are there to move you around.

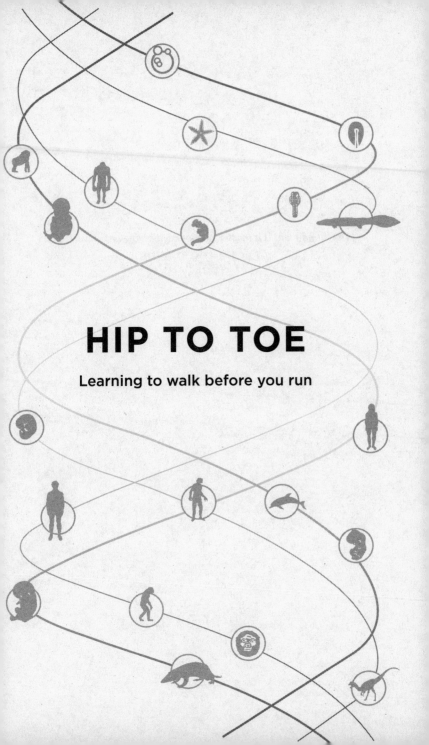

HIP TO TOE

Learning to walk before you run

'The human foot is a masterpiece of engineering and a work of art'
LEONARDO DA VINCI

TWO LEGS GOOD

Most of our tetrapod cousins, whether reptiles, amphibians or mammals, have stayed faithful to the original mode of locomotion – getting around, on land, on four legs. But not all tetrapods are quadrupeds, there are, of course, quite a few who have done something very different. Some tetrapods have given up a terrestrial existence entirely, favouring a return to a more watery way of life. Whales, dolphins and porpoises are all tetrapods, thanks to their evolutionary heritage, but they are no longer four-legged land animals. Other tetrapods have evolved to be able to lift themselves up from an earth-bound existence, using their limbs to move through the air as easily as a fish moves through water. These ethereal creatures are the descendants of reptiles, and in fact, more precisely than that, they are the descendants of dinosaurs. Really we should class them as dinosaurs, which means of course that not all the dinosaurs died out 66 million years ago: some survived and thrived and are still with us today – these are the feathered, winged dinosaurs which we know better as birds. (Next time you get up close to a bird, whether it's a robin in the garden, a seagull on a quay or a chicken on a farm, look deep into its inscrutable eye. That's a dinosaur you're looking at!) Other tetrapods are still land-bound but have lost their limbs entirely: snakes had four-legged ancestors just like you or me, and still others walk around on land on just two of their legs. That includes birds, of course, who tuck their feathered forelimbs away when they deign to walk on the ground. But there are also a few mammalian bipeds, including kangaroos and wallabies, kangaroo mice, jerboas, and *us*.

If you had to sum up the crucial difference between us and other apes in just two words, they would be: brains and bipedalism. Expanding brain size is a major theme in human evolution, but long before brains started to grow bigger, our ancestors began to walk *habitually* on two legs – and this is a major difference between us and any other living ape. It's very important that we make the distinction of *habitual* bipedalism, though, as all the other apes are capable of walking around on two legs, but they don't do it routinely like us.

Standing, walking and running upright on two legs means that the shape and size of our legs – and the pelvis, which connects our lower limbs to our spines – is very different from that of other apes: humans today have a basin-shaped pelvis, relatively large hip and knee joints, an inward angle at the knee (we're all a little knock-kneed), a differently shaped ankle joint, a springy foot, shortened toes and a big toe brought into line with the other toes. These features didn't all arrive in one fell swoop, and it would be very odd if they had done.

This brings us to thinking about the forces that shape our particular anatomy. What has sculpted your body into its precise shape? Your DNA plays a leading role, of course, but even genes and the other bits of DNA that influence them don't produce a precise blueprint of a body. They direct the way in which the body develops, and they effectively set parameters within which anatomy may vary. The shape and size of various features will settle, within those parameters, depending on environmental influences. An obvious example is nutrition. Imagine two genetically identical twins, one growing up on a poor diet and the other well nourished. You wouldn't be surprised if the second twin ended up bigger and taller than the first. In fact, it's not possible to draw a clear line between genetic and environmental influences on development and the structure and function of the body. Fairly recent research has revealed that some environmental influences work via the genes, which become chemically modified so that their function is altered. It's also recently been discovered that these modifications to genes can be inherited along with the genes themselves. The study of these changes is called epigenetics (not to be confused with the concept of embryological development involving complex tissues differentiating out of simpler ones: epigenesis). Beyond what might be considered 'normal' influences on anatomy, there's another force at play: pathology. This covers a multitude of sins, as it were: problems in DNA, which might arise from copying errors, which lead to congenital defects, problems with metabolism or cancer; infections like bacteria and viruses; malnutrition, whether that's a general lack of food or a lack of specific components like vitamins; and physical trauma.

Focusing on what you might call variation in 'normal anatomy' (excluding pathology), it's clear that the features that we now consider to be unique to humans among the living apes did not appear in a flash, they arrived in a piecemeal or mosaic fashion. The sprig of the tree of life which includes us and those ancient species which are *our* ancestors, but not those of any other living ape, as far as we know, sprouted about six to seven million years ago. All the twenty or so species represented on this sprig are hominins. This has a very precise meaning in biological classification. We belong to a wider family, Hominidae (ie: hominids or great apes), which includes us, chimps, gorillas, orangutans and all of our ancestors. But within the hominid family there's a subfamily of Homininae (gorillas, chimps and humans), and within that, a tribe called Hominini (hominins) – an exclusive club of humans and our own ancestors. Our earliest hominin ancestors have adaptations to fairly habitual bipedalism, in fact, that's exactly why we presume they're our own ancestors, rather than those of our closest living relatives, chimpanzees.

In 2001, a team led by the French palaeontologist Michel Brunet discovered a distorted, partial fossil skull in Chad, which was dated to 6–7 million years ago. It was a new species, named *Sahelanthropus tchadensis*, but it also became known by the nickname 'Toumai', which means 'hope of life' in the local Goran language. The braincase was tiny but there were a few intriguing characteristics which suggested that this fossil skull might belong to one of our hominin ancestors. Compared with modern chimpanzees, whose jaws project forwards, Toumai has a flat face. But underneath this skull, there could be an even more important clue, in the position of the foramen magnum, the large hole in the base of the skull where the brainstem exits to become the spinal cord. In chimpanzee skulls, this hole is situated towards the back of the skull. Toumai's skull is badly distorted, but it has been suggested that the foramen magnum was further forwards, and that this skull would have therefore been balanced on an upright spine. It sounds like a subtle distinction, but this connection between the position of the foramen magnum and an upright spine holds true for other animals too. Bipedal kangaroo rats have a more anteriorly positioned foramen magnum than quadrupedal mice, and the same is true for bipedal wallabies compared with quadrupedal tree kangaroos. Unfortunately, there's a lot of variation in the position of the foramen magnum under both human and chimpanzee skulls, which means it's impossible to be sure whether Toumai really was an upright, bipedal ape: a hominin.

Palaeoanthropology is full of debate – it's one of the reasons why it's such an interesting discipline. One of the big problems with Toumai's skull is that it is *just* a skull. Although there have been rumours that a femur was discovered at the same time, this bone has never been described in a publication. While the position of Toumai's foramen magnum may

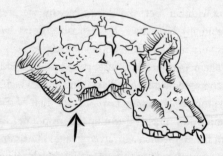

Toumai – with the position of the foramen magnum indicated

suggest that he walked upright, it's very difficult to be sure without any evidence from the rest of his skeleton. Not everyone agrees with Brunet's interpretation of Toumai's skull: some researchers have suggested that Toumai isn't a hominin at all, and that he could instead be a common ancestor of chimpanzees and humans. Or even an ancestral gorilla.

LUCY'S HIPS AND JOHANSON'S KNEE

With later fossils, we have much more of the skeleton to go on, and one of the best-preserved and most famous hominin skeletons is 'Lucy', who was discovered by Don Johanson in 1974. Johanson was part of a team of American, British and French anthropologists who were working in the Afar region in northern Ethiopia, surveying the Hadar Formation, which contained fossil-bearing sediments dating to around three million years ago.

I was lucky enough to meet Lucy, in the flesh, while making the series *Prehistoric Autopsy* for BBC2. One programme focused on her and her species, *Australopithecus afarensis* (the 'southern-ape from Afar'), and as well as having an accurate cast of her skeleton in the studio, we had commissioned a life-size reconstruction which we unveiled at the end of the programme.

Lucy is absolutely tiny, only standing a metre tall, and there's absolutely no doubt with her species that we're looking at a hominin, an ape which was habitually bipedal on the ground. The shape of her hips, particularly, is very human-like – completely different from those of a chimpanzee. The upper bone of the pelvis, the ilium, is short and broad in hominins, compared with the tall and narrow ilium of chimpanzees.

The broad ilium of hominins is very important when it comes to positioning some key muscles around the hip joint. Three gluteal muscles attach from the outer surface of the ilium and insert into the neck of the femur: gluteus maximus, medius and minimus. Gluteus maximus, a very large muscle, as its name suggests, overlies the other

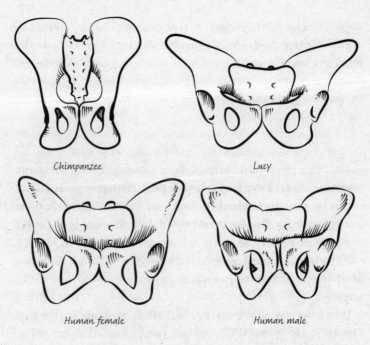

Chimpanzee

Lucy

Human female

Human male

The hips of Lucy, a chimpanzee and humans

two. It's a powerful extensor of the hip joint, pulling the thigh backwards in relation to the pelvis. You use this muscle to straighten your flexed legs underneath you when you stand up from sitting or when you climb stairs. The other two gluteal muscles attach from the side of the pelvis down to a lever which projects from the neck of the femur called the greater trochanter. Rather than extending the hip joint, these muscles act to move the thigh out to the side, or 'abduct' it. But their real importance to us habitually bipedal hominins is what they do when they're contracting to stabilise the hip joint, without actually moving it. This happens every time you take a step. As you lift a foot off the ground and swing your leg through, you're left standing on one leg. At this moment, there's a tendency for your pelvis to drop down on the unsupported side – if this happened as you tried to swing the leg forwards to take that step, you'd end up dragging your foot along the ground. But the hip abductor muscles (gluteus medius and minimis) on the leg in stance contract to pull the pelvis down to the femur on that side, and stop it dipping down to the opposite side.

The importance of this arrangement of muscles around the hip joint is revealed when someone *loses* the use of their gluteus medius muscles, which can happen if the nerve supplying it is damaged. As a patient with this problem takes a step forwards, their pelvis, unsupported by gluteus medius on the stance side, dips down over the moving leg. In order to compensate for this, the patient will pull their whole body further over to the side of the leg in stance. The result is a peculiar, lurching way of walking known as 'gluteus medius lurch' or 'Trendelenburg gait'. Friedrich Trendelenburg was a pioneering German surgeon who lived from 1844 to 1924 and whose name is immortalised in a range of surgical problems, tests and treatments, including that abnormal gait, as well as a test and an operation for varicose veins.

Lucy, as her wide pelvis shows, would not have lurched around with a Trendelenburg gait, and neither, it seems, would she have walked with the bent-hip, bent-knee gait of a chimpanzee. Her lumbar spine would have been bent backwards into a lordosis, like yours, to balance the weight of her trunk over her pelvis.

There are further indications that Lucy would have walked with quite straight legs from her knees. In fact, the very first fragments of fossil skeleton that were named *Australopithecus afarensis* were the lower end of a femur and the upper end of a tibia. Discovered a year before Lucy herself was found, and by the same palaeoanthropologist, these fossils quickly became known as 'Johanson's knee'. Crucially, they show that the femur and the tibia meet at an angle, and this is another telltale sign of habitual bipedalism. With the fossils of just that ancient knee, Johanson knew that he was looking at a bipedal ape: a hominin. In quadrupeds, the femur tends to be straight, with no angulation at the knee joint, but

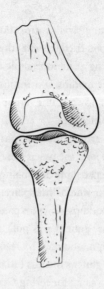

Johanson's knee

in humans and their bipedal ancestors, the femur slopes inwards, down towards the knee, creating an angle. Even if you don't think of yourself as knock-kneed, you are a bit, especially in comparison with your chimp cousins. As you walk along, you are balancing your body weight over one foot, and then the other, and to do this you need to be able to put your foot right underneath your centre of mass. Your sloping femur and angled knee help you to do this.

The bicondylar angle is only useful if you walk with straight legs, so the very presence of it in a skeleton suggests just that: australopithecines, like us, walked with straight, not bent knees. It's much more efficient to walk with straight legs – try walking around with your hips and knees bent for a bit, and you'll realise how hard your muscles, particularly your quadriceps at the front of your thighs, are having to work.

Although the bicondylar angle is so useful to efficient bipedal walking, it produces a knock-on problem for knock-knees. The large four-headed quadriceps which make up the bulk of the muscle on the front of your thigh crosses the knee joint, where the kneecap or patella is embedded in its tendon, and inserts onto a bump on the front of the tibia. Because your femur is angled, so is your quadriceps muscle, and its line of action is not through the centre of the knee joint but a little off to the side. This means that, when you contract your quads to straighten your knee, the patella is also being pulled sideways a bit – towards the lateral or outer side of your knee. Your anatomy compensates for this (it has to, otherwise your patella (kneecap) would be dislocating sideways all the time): there's a steep shelf on the outer lip of the groove on the lower femur, where the patella sits, and the attachment of the fleshy fibres of the medial head of quadriceps extends further down than the lateral head, meaning that the medial part of this muscle will help to pull the patella towards the middle and stop it dislocating. There is also strong fibrous tissue attaching to the patella to help keep it in place.

Moving down to the ankle and foot, there are more indications that Lucy's kind were habitually walking around on two legs. Ankle joints are hinge joints, formed by the tibia and its companion bone in the lower leg, the fibula, and the uppermost bone of the foot: the talus. Most of the loading through the joint occurs between the talus and the tibia, with the fibula bracing the joint. The lower tibia and fibula are firmly bound together, helping to stabilise the ankle joint. Both bones form projections which extend down on either side of the talus, gripping it and holding it in place. If you look down at your foot and pull your toes up towards you, this is dorsiflexion. Now point your toes down and you've plantarflexed your foot. Both these movements occur at the hinge that is the ankle joint. You can also bring the sole of your foot to face inwards (this is called inversion) or twist it outwards (in eversion). These movements *don't* happen at the ankle joint: they occur further down, in the foot itself.

The anatomy of our ankles and feet mean that the soles of our feet rest squarely on the ground. Chimpanzee ankles are different from ours, with an 'inverted set' – in other words, their ankles naturally rest with the soles of the feet slightly facing in towards each other. The articular surface of the tibia, where it forms the joint with the talus, is perpendicular to the axis of the tibia itself in humans. In chimpanzees, it's angled, producing that natural inversion of the foot. Chimpanzee ankles also allow the foot to flex right back, forming an angle of around 45 degrees with the shin. This enables them to climb vertical tree trunks with ease. Gorillas, orangutans and gibbons are all similar to chimpanzees in this regard. In contrast, from its resting position perpendicular to your shin, you can probably bend your foot up (the technical term for this is dorsiflexion) just 15 degrees, until it forms an angle of about 75 degrees with the shin. Pushing it to 45 degrees – something which is entirely normal in other apes – would be likely to injure your ankle.

Human feet look very different from those of other apes. Look at a chimpanzee's foot and it's clear that this is an organ which is well designed for grasping. The foot is flexible and the toes are long, and the big toe sticks out to the side rather like a thumb. Your foot looks very different; it's more rigid, with an 'instep' (which, as an anatomist, I'd call the medial longitudinal arch), formed by the natural shape of the bones in your foot and supported by ligaments and tendons. There's a less obvious arch on the outer, lateral edge of your foot, as well as a transverse arch, running side to side across the width of the foot.

These arches make the foot into a springy platform. As you walk or run, the elasticity of the ligaments and tendons supporting those arches enables you to conserve a little energy each time you take a step – the tendons are stretched a bit as your foot touches down, and spring back as your foot leaves the ground. Your toes are relatively short compared with those of your chimpanzee cousins and your big toe is neatly lined up with the other toes. This is a foot for walking and

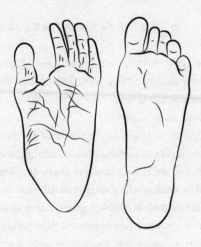

A chimpanzee and a human foot

running on, not for grasping things with. (Although your foot still has all the muscles that are present in the chimpanzee foot, so it's still possible to develop that grasping potential to some extent. People who have lost the use of their hands, or who have been born without hands, show us how it's possible to rediscover that prehensile capability in our feet.)

In the year 2000, an extremely well-preserved foot bone was discovered in Hadar, in Ethiopia. Over the course of 25 years, more than 250 fossils have been recovered from the particular site where the foot bone was found, all of them dating to around 3.2 million years ago, and all of them appearing to belong to members of Lucy's species, *Australopithecus afarensis*. It's reasonable to assume that this foot bone belongs to the same species. To be more precise, this bone is a left fourth metatarsal; it's an important bone because it has a twist along its length that shows it was part of a human-like foot with a transverse arch – like your foot, rather than part of a chimpanzee-like, flat foot.

ANCIENT FOOTPRINTS

Fossilised bones aren't the only reason why we believe that australo-pithecines regularly walked on two feet much like us. The fossils would be enough, certainly, but there's some other evidence from Tanzania which is difficult to argue with.

Some 3.7 million years ago, a volcano called Sadiman erupted in what's now Tanzania, ejecting vast clouds of ash into the sky. It was the wet season, and after the clouds of ash settled, rain fell and turned the ash layer into a fine, sticky mud. Animals walked across it, leaving trails of their footprints behind: elephants, giraffes, antelopes, guinea fowl, and some of our ancestors. Three ancient people walked across the ash, leaving a trail of footprints which, against the odds, would be preserved for millions of years. In 2011, I travelled to Tanzania and met a man who now runs safari camps for tourists and television crews. We were making a programme about diet in human evolution, and hoping to film with a group of Hadza hunter-gatherers, deep in the bush. The safari organiser's name was Peter Jones, and I knew I'd heard his name before.

Back in 1977, before he started organising safaris, Peter had been working with Mary Leakey's team at Laetoli in Tanzania, looking for hominin fossils, and he had found what looked like human footprints in the ancient ash layer. It was an amazing discovery which grabbed head-lines when Mary Leakey spoke at press conferences in the US later that year. The Laetoli footprints showed, beyond a shadow of a doubt, that almost four million years ago hominins were walking upright.

Careful analysis of the footprints by Robin Crompton at Liverpool University – using laser scanning to measure depth and calculate pressure across each foot as it squelched into the sticky ash – has revealed *how* the Laetoli hominins walked. It is likely that those hominins belonged to the species whose fossils have been found nearby: Lucy's kind, *Australopithecus afarensis*. The footprints in the ash provide an independent line of evidence and corroborate inferences drawn from the fossilised skeletons of these ancient people. Once again, the footprints imply that these hominins were walking with straight legs like us, rather than with bent hips and bent knees like chimpanzees.

THE LIMITATIONS OF HUMAN FEET

But when did our ancestors' ankles and feet change? To answer this question, we need to cast our net more widely and look at more distant living relatives and even earlier hominin ancestors. We also need to be really sure about the ranges of both anatomy *and* function in living apes, including ourselves. Human locomotion is perhaps more varied than we might imagine from the perspective of our own lifestyles.

Firstly: ankles. We've noted that human feet don't bend up as far as chimpanzee feet. A quick test convinces me that this is true for my feet, and it probably is for yours too. But, like me, you've grown up in a country where we don't tend to climb trees much. Our anatomy isn't just shaped by genes, but by environment too. Just how different would our ankle joints be if we had been climbing up vertical tree trunks from a young age?

Some modern humans do much more tree-climbing than I bet you've done over the course of your life. There are modern hunter-gatherer societies where tree-climbing is a regular occurrence in order to obtain certain fruits and game, and – especially – honey. Climbing trees up to dizzying heights of more than 50 metres to reach honey is well documented among the Mbuti and Efe people from the Ituri forest of the Democratic Republic of Congo, and the Batek of Malaysia. Climbing trees this high is a dangerous activity, but worth it for the honey itself and the prestige which goes with it. Sometimes, climbing equipment is used, especially if a tree trunk is very wide, but on thinner trunks the honey-hunters will climb up unassisted.

So exactly how much ankle dorsiflexion goes on in this type of vertical climbing? Anthropologists recently filmed Twa hunter-gatherers in Uganda to answer that question precisely. Twa men climbing trees to gather honey had considerably more flexibility in their ankle joints than I do. They would dorsiflex their ankles to even more than 45 degrees as they climbed. This extra flexibility at the ankle joint appears to come down to soft tissue anatomy, particularly the stiffness of the calf muscles, rather than to the bony configuration of the ankle joint. If living humans are capable of climbing up vertical tree trunks in this way, early hominins like Lucy probably could have done it too. Our anatomy is more flexible than we sometimes imagine, especially when we do something regularly and from a young age.

So what about our feet below the ankles? There seem to be such obvious differences here between the feet of humans and other apes, and it also seems obvious to assume that chimpanzee feet, for example, have retained an ancestral flexibility while our feet have become stiffer. We have arches in our feet, and strong ligaments in the soles of our feet that are missing in other apes.

If we look beyond apes, at the feet of monkeys, we find something rather interesting. Monkey feet are not all that flexible; they are fairly rigid, which is important for leaping: it would be more difficult and less efficient to push off from a support with a more flexible foot acting as a lever. But African apes are too large to safely leap around in trees like smaller monkeys, and apes can get away with having more flexible feet. As we've seen, their feet are indeed flexible, very good at grasping, and appear almost hand-like. For a long time there's been an implicit assumption that hominin feet must have started off looking much like a chimpanzee foot, but the stiffness of monkey feet suggests there are other possibilities. It seems reasonable to assume that very ancient ape feet might also have been stiffer.

The discovery of a set of fossils belonging to a species called *Ardipithecus ramidus*, dating to around 4.4 million years ago, forced palaeoanthropologists to look at hominin feet (and many other bits of the hominin skeleton) in a new light. 'Ardi', as she came to be known, has feet more like those of a monkey than an African ape. Unlike other hominins, Ardi's big toe stuck out to the side: good for grasping branches but not so good for pushing off from the ground during walking. But Ardi also had a stiff foot – more like those in monkeys and in modern humans.

But just how stiff are human feet compared with other apes'? Interpreting fossil footprints depends on a really good understanding of form and function in living people (and other primates). For more than 70 years it's been assumed that human feet are very stiff and don't bend in the middle like other apes' feet. But a team of scientists – including Robin Crompton, who has always been fascinated by feet and footprints – recently put modern human feet to the test. Using a pressure-recording treadmill they were able to collect data about the pressure changes under people's feet as they walked, and they uncovered a surprising range of variation. Just as human ankles seem to be more varied in their function than we've previously assumed, it seems that we've also been too rigid about the stiffness of human feet. Some people had an 'ape-like' flexible mid-foot with a mid-tarsal break.

Another team of anthropologists collected data from 398 people walking over a pressure-sensitive pad and found that 8 per cent of them had the same mid-tarsal break. It was also clear that this wasn't an 'all-or-nothing' feature; participants in the study showed a range of flexibility in their feet. The people with more flexible feet also tended to be flat-footed, with low arches.

Robin Crompton's team found that there was also variation *within* individual people: a foot could change in its flexibility. In fact, an incredible two-thirds of the people they analysed showed evidence of a mid-tarsal break in *both* feet within five minutes of walking on their treadmill. This suggests that the main factors controlling stiffness are not bones but soft tissues like ligaments, tendons and muscles. Once again, this means we have to be cautious when drawing inferences about function from fossil bones. We need to understand variation in the anatomy and function of living people's feet first. The fact that each of our feet can vary in its stiffness is really important in itself: this may help us to walk on different types of surfaces.

Varying pressures under a foot during walking (from Bates et al. 2013)

The researchers also collected data from bonobos and an orangutan, and found something perturbing. The results overlapped with the human data. Although the non-human apes typically showed high pressures along the outer edge of their feet, so did some humans.

The variation that's now being picked up in modern human feet (which increases if we include people who haven't grown up wearing shoes) makes it very difficult to interpret function in fossil feet and to infer how much time a particular hominin was spending walking on the ground rather than in trees. It's now clear that feet that look really quite different can end up being functionally very similar, and particular features that seem so obviously linked to a particular function – like an opposable big toe and grasping tree branches – aren't that easy to interpret. As an example, the lowland gorilla is the most arboreal African ape, but it also has the most human-like feet.

The function of feet appears to be more variable than we've thought in the past, but this doesn't mean we should give up – clearly, much more work needs to be done. If we're ever going to really understand how a fossil species moved around and interacted with its environment, we need to make sure we *really* understand how the anatomy of living species is related to function, and how both form and function are related to ecology.

We also need to remember that any living species is unlikely to closely resemble an ancient ancestor. The anatomy of hominins has changed considerably over the five to seven million years since the last common ancestor with chimpanzees, but anatomy within the chimpanzee lineage has undoubtedly changed too. This leaves us wondering just how chimpanzee-like was that last common ancestor? Tied up with this question is perhaps an even bigger one: where did human bipedalism come from?

THE ORIGINS OF BIPEDALISM

The most famous illustration of human evolution must be the 'March of Progress'. You know the one, it starts with a chimpanzee-like ancestor on the left and progresses through a series of gradually more upright-looking creatures until we reach the apogee with a thoroughly modern man on the right. There are all sorts of spoofs on this theme – google 'march of progress' and you'll soon find them. There's a great one where the man starts to hunch over again, carrying a rake, then a pneumatic drill, and ends up with a man sitting in front of a computer. There's one with a Dalek evolving from a vacuum cleaner, and a Simpsons' one ending with 'Homersapien'. Someone's even made and photographed one built of Lego.

This iconic image has firmly embedded itself in our culture and, I would argue, in our brains. But it's ended up emphasising and perpetuating two very unhelpful ideas about human evolution. Firstly, it has helped to encourage the idea of human evolution as an unflinchingly linear progression, perhaps even a predestined 'March of Progress'. Evolution, including within our small sprig of the tree of life, is not a ladder of progress but, to quote Stephen Jay Gould, a 'copiously branching bush'. The original illustration of 'The Road to *Homo sapiens*' was drawn by Rudolph Zallinger for a popular science book by the American anthropologist F. Clark Howell, entitled *Early Man* and published in 1965. Howell didn't describe a linear trajectory of hominin evolution in his text, and the illustrations – which are reconstructions of fossil ape and hominin species known at the time – are simply arranged in a broadly chronological order. It's clearly not intended to imply that one species evolved into another, all the way along the chain, but this is the power of images – from a cursory glance at the illustration, it *looks* that way.

The image helps to create another, potentially unhelpful, impression. It appears to show that bipedal hominins evolved from a chimpanzee-like, knuckle-walking ancestor. This is perhaps slightly unfair to the original image, which includes drawings of fifteen species, and the first two – *Pliopithecus* and *Proconsul* – are actually shown walking upright. The full illustration continues overleaf, with a fold-out page starting with a quadrupedal *Dryopithecus* (a Miocene ape from Europe and Asia). Many of the numerous re-drawings of this image reduce it to five or six species, and start with the knuckle-walking *Dryopithecus*, who is drawn to look very much like a modern chimpanzee. So here's what I think is the other problem with this image: it has crystallised the idea in our brains that bipedal humans like us have evolved from a chimpanzee-like, knuckle-walking ancestor. In recent years, the revelation that we are genetically very similar to chimpanzees (it depends how you measure it, but sometimes the similarity is quoted as being as much as 99 per cent the same), has only helped to promote the idea of a chimpanzee-like ancestor of humans. But this really is a big old assumption.

Bipedalism is considered to be a fundamentally human character-istic, but we should be rather careful and distinguish *habitual* bipedalism as our special characteristic, as all living apes are capable of walking around on two legs – they just don't tend to do it as much as us. For you and me, walking (or running) on two legs tends to be our 'locomotor mode' of choice. Quadrupedalism is for babies; climbing and hanging by our arms is for children and some more adventurous adults. (This is not to say that we aren't very good at climbing, we are, and we'll explore that more in a few pages' time.)

The origins of (habitual) bipedalism in humans has been the subject of intense debate. In 1863, four years after Darwin published *On the*

Origin of Species, his 'Bulldog', Thomas Henry Huxley, published his own account of human evolution, *Evidence as to Man's Place in Nature*. The frontispiece showed a series of skeletons: gibbon, orangutan, chimpanzee, gorilla and human, all posed in an upright position. It seemed obvious that humans had evolved from apes who were already upright in the trunk, thanks to hanging around by the arms (or brachiating) in trees. This continued to be the prevailing view into the early twentieth century: that at some point our tree-living ancestors dropped out of the trees and continued to move around with an erect spine, but this time walking on their two feet. The Scottish anatomist, Sir Arthur Keith, conservator of the Hunterian Museum at the Royal College of Surgeons in London, encapsulated this idea in his 'hylobatian theory' in 1923. Hylobatids are gibbons, and these apes have perfected the art of arm-swinging – they're quite amazing to watch if you ever get the chance – but when they drop down to the ground, they walk on two legs.

But ideas about the origins of human bipedalism shifted in the later twentieth century. Gibbons were forgotten as the new science of genetics revealed that humans were closely related to the African apes: chimpanzees and gorillas. Those apes tend to move around on the ground on all fours, walking on the soles of their feet and the knuckles of their hands, or 'knuckle-walking'. So it seems very reasonable to assume – and this would be the simplest, most parsimonious evolutionary path for all concerned – that humans, chimpanzees and gorillas have all evolved from a knuckle-walking common ancestor. Indeed, researchers have identified bony features in the hands and wrists of apes and modern humans which look as though they might be adaptations to knuckle-walking. (The idea being that we still have these features because we once had a knuckle-walking ancestor, even though we don't indulge in it any more.)

It pays to be cautious about linking particular bony features to specific functions without extremely good evidence to back this up. In fact, it pays to be cautious in many areas of science. We formulate hypotheses based on what we already know, but a good hypothesis is one that has withstood being tested. As we gather more data, we can expect that some hypotheses won't stand the test of time, and we might have to come up with new explanations for the patterns we see. This doesn't mean that we can't trust scientific theories (which, after all, are the same thing as scientific 'facts'), it just means we have to be open-minded and accept that explanations may change. Perhaps this makes the whole edifice of science seem rather shaky, but it shouldn't. There are some theories, some facts that are so unlikely to be disproved that we can depend on them. So, for instance, that the Earth really is spherical, not flat, and that evolution has definitely happened. But when it comes to how our ancestors moved around, before they started habitually walking on two legs on the ground, there's more room for debate and there are now some significant challenges to the whole concept of a knuckle-walking ancestor.

Firstly, the elegance of the suggestion of a knuckle-walking ancestor for chimpanzees, gorillas and us only works if the knuckle-walking of chimpanzees and gorillas is essentially the same, or homologous. Several studies over the last fifteen years or so have called that assumption into question, showing that knuckle-walking is subtly different in chimpanzees and gorillas, and that their wrists also develop in different ways. Some of the bony features which have been assumed to be related to knuckle-walking have also been questioned. Some relate to stabilising the wrist joint, but that might be something which evolved principally for vertical climbing and just happened to be useful for knuckle-walking too.

The anatomy of Ardi (*Ardipithecus ramidus*, first discovered in the early 1990s) also makes it seem less likely that humans ever had a knuckle-walking ancestor. Ardi is a very ancient hominin: at 4.4 million years old we're getting pretty close to the date of our last common ancestor with chimpanzees, given that genetic comparisons estimate that our paths diverged some five to seven million years ago. There are plenty of fossil bones of this species to look at, and what those bones reveal is that Ardi was not much like a chimpanzee. If anything, she was more like a big monkey, rather like the earliest fossil apes, including species like *Proconsul*. We've already seen that Ardi had stiff feet – but an opposable big toe. This combination of features in the foot reflects the flavour of the skeleton more generally: Ardi combines what look like adaptations to a tree-living existence with characteristics which suggest she was also a fairly regular biped on the ground.

What Ardi doesn't have is any features which suggest she knuckle-walked, or indeed any features which would suggest she did a lot of hanging from her arms. Rather than having a stiff wrist like non-human apes today, she had a very flexible wrist, which could bend backwards a long way. Although she's a larger animal, Ardi's limb proportions are similar to those of modern macaques, and it's been suggested that Ardi would have moved around in the trees rather like a macaque, on all fours, using the soles of her feet and the palms of her hands. However, Robin Crompton and his colleagues have pointed out that Ardi, at around 50kg, is very large compared with macaques – more than five times larger, in fact. Her large body size and stiff feet (and, as an ape, her lack of a tail as well) mean that she would have had trouble balancing on branches on all fours. Instead, they suggest that Ardi was doing something that all apes do, something that doesn't make walking on the ground seem like such a giant leap: Ardi was walking in the trees.

It makes a lot of sense – standing on two legs, Ardi could have used her hands to support her, gripping on branches above or to the sides, or pushing down, using her palms and fingers, on branches below shoulder height. In this way, a fairly butch ape could quite safely move around, high up in trees. In fact, all great apes still use this way of moving around in trees: walking on two feet along horizontal boughs, using their hands to help them balance. Among all the great apes, the orangutan is the one who spends the most time up in the trees – and also happens to be the most bipedal. When orangutans walk on two legs, they tend to do it with straighter hips and knees than chimpanzees and gorillas, who both tend to use that more flexed and inefficient bent-hip, bent-knee gait.

It's interesting to look at *when* orangutans do their 'walking in the trees'. Although they're bipedal for less than 10 per cent of the time they spend moving in trees, it's important to them. Standing on branches on two legs means that they can get right out to the edges of trees, to reach fruits there, and to bridge across to the next tree – much easier and more efficient than getting down from one tree and climbing up another one. As well as moving along horizontal boughs, orangutans also stand on branches when they're climbing up. This type of climbing seems much more familiar to us, I think, than the climbing up vertical tree trunks that comes so naturally to chimpanzees and gorillas (although we've seen that honey-hungry foragers seem to be able to do that quite easily, too).

For me, this walking and climbing in the trees, using horizontal branches where possible, makes human bipedalism seem like a much more natural way of moving. Instead of a brand-new form of locomotion, it's something that was already there – probably going way back in ape evolution. While there's been plenty of heated debate over *when* and *how* bipedalism arose, there's also been well over a century of head-scratching about *why* it appeared. Darwin suggested it allowed the hands to be used for defence, while others have suggested that it freed up hands for carrying helpless babies or food. Or it allowed hominins to see over the tall grass of the savannah and spot predators. Or, on the hot savannah, standing on two feet meant that less of the body was exposed to the scorching sun. Or it originated as an aggressive posture, or from an aquatic phase when our ancestors spent a lot of time wading in water. Or it's just more efficient to move on two feet rather than four. Or it allowed males to impress females with their conspicuous external genitalia. Or it was a gimmick which caught on. Or a combination of the above.

There are flaws with many of these suggestions. The babies of early hominins probably wouldn't have been helpless, as brain size expanded much later than the appearance of habitual bipedalism in hominin evolution. Chimpanzees manage quite well to carry food while moving along on one arm and two legs. Explaining bipedalism as an adaptation to a savannah habitat doesn't work as we know now from reconstruction of ancient environments that the grasslands underwent a major expansion from around three million years ago, long after hominins had started walking around on two legs as a matter of course. There are too many problems with the strangely enduring aquatic ape hypothesis to discuss them here, but it's sufficient to say that there is no evidence at all for any even semi-aquatic phase in hominin evolution. Anyway, the point is this: if terrestrial bipedalism is not such an extraordinary thing for an ape to do, and in fact really quite ordinary, we no longer need to come up with an extraordinary reason for doing it. If terrestrial bipedalism evolved in knuckle-walking apes, then you do need to explain why there was a sudden 'flip' from one mode of locomotion to another. But if you view habitual bipedalism as the expansion of one particular mode of locomotion which was *already there* in an ancestor, the transition is much smoother. For apes making use of more open habitats, rather than dense forest, it was a natural move, made easier by the fact that ape ancestors were already bipedal above the ground.

In fact, walking in the trees seems to have been important to several ancient species of ape, including *Pierolapithecus, Hispanopithecus* and *Morotopithecus*, which lived in the Miocene epoch (between 5.3 and 23 million years ago). It seems that at least one Miocene ape was also spending enough time walking on the ground for that to shape their anatomy. *Oreopithecus bambolii* is an ancient ape which lived between seven and nine million years ago, in what is now Tuscany and Sardinia. *Oreopithecus* possesses some familiar features: a bicondylar angle at the knee, and hips and knees apparently adapted for use in extension: this certainly looks like an ape which walked on the ground. Ardi, who lived some five million years later, may have moved around in a similar way, combining walking in trees with walking on the ground.

Having criticised the image of a knuckle-walking ancestor which has been so successfully marketed by the edited-down version of Zallinger's 'March of Progress', I hope I'm managing to erase that persistent picture of a knuckle-walking ape, standing up and walking, from your mind. There's now very good evidence that our hominin ancestors never walked on their knuckles but instead were tree-walkers who became ground-walkers. This idea of the origins of bipedalism, even terrestrial bipedalism, pushes it back way before the first hominins arrived on the scene, back to the ancient Miocene apes who were the ancestors of orangutans, chimpanzees and gorillas, as well as us.

It challenges our tendency to think of ourselves as the special ones, but if this version of events is true, then it's chimpanzees and gorillas that have been the innovators, with their ancestors independently coming up with knuckle-walking as a new way of getting around. In fact that's probably not the only innovation they came up with in parallel. Looking at the forelimbs (and indeed, the hindlimbs) of living and fossil apes, it seems most likely that many adaptations to climbing and hanging from arms, including stabilising features at the wrist joint, have evolved several times in parallel. This of course suggests that climbing and arm-hanging themselves have evolved several times. This might appear to be an unnecessarily tortuous version of events (wouldn't it be simpler, more parsimonious, if this behaviour and these features just arose once?) but the fossil record doesn't show a steady progression – it's much more piecemeal. It also makes more sense if you consider that climbing and arm-hanging are ways of moving which make sense as soon as you reach a certain body size. (Remember the argument about Ardi being a quadruped in the trees? It's difficult to balance a large body in that way, and much easier to stand or climb, bearing your weight on your legs, using hands to support you, or, as some apes do, to hang from branches.)

Among these innovative and inventive apes, then, it seems that we humans are the conservative ones; just doing what apes have always done, walking around on our two feet.

BORN TO RUN

Our ancestors have been walking around on two legs for much longer than we've previously thought, if we take into account walking in the trees as well as walking on the ground. But when did they start to look like us in terms of their general body plan, their limb proportions? Ardi, and even australopithecines, look very much like other apes, with their long arms and short legs. (Until 2005, Lucy was the only *Australopithecus afarensis* skeleton that included elements of both arms and legs. Then, an Ethiopian team discovered another partial skeleton in the Afar region – dated to 3.6 million years ago and called Kadanuumuu (Big Man) – much larger, probably a male. He's tall but his legs are still short compared with modern humans.) Long legs, which are so characteristic of humans today, appeared relatively late in our evolution.

The first evidence of a truly *human* body plan, complete with long legs, came to light in 1984. Kamoya Kimeu, while helping to lead Richard Leakey's fossil-finding team, discovered the remains of a fossilised skeleton beside the sandy bed of the Nariokotome River, 3 miles west of Lake Turkana. It was the skeleton of a boy – his bones were not yet fully grown – and he's become known as 'Nariokotome Boy'. He lived in East Africa 1.5 million years ago and he belongs to a species known as *Homo erectus*. Some researchers separate out early, African *Homo erectus*, placing them (including our Boy) in a species called *Homo ergaster*. But support for this extra species has waned, especially since *Homo erectus* fossils from just one site, Dmanisi, in the Republic of Georgia, has shown this species to include a wide range of variation.

There are no obvious signs of the cause of his death on the skeleton of Nariokotome Boy, although some have suggested that a tooth abscess may have finished him off. That's possible, certainly – with no antibiotics to help combat the infection – but not at all definite. However he died, his body must have been quite quickly covered up with sediment in order for it to have escaped from the teeth of scavengers and been so well preserved.

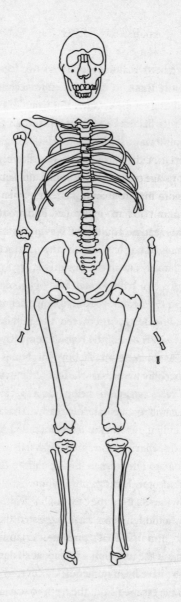

Nariokotome Boy

The age of Nariokotome Boy at his death has been the subject of considerable discussion over the years. From his bones, which show signs of unfused growth plates, he looks like he should be a teenager. At least, if he were a modern human, I'd say he was somewhere between ten and fifteen years old. But his teeth tell a different story. The mismatch in the age indicated by his bones and by his teeth suggests that he didn't develop in the same way as you or me. He didn't develop in the same way as chimpanzees, either. Anyway, the most recent, best estimate of Nariokotome Boy's age puts him very young, only about eight years old. But he's much more mature than a modern human eight-year-old. He probably stood about 154cm or 5'1" tall, and would have been three inches taller had he reached adulthood. Previous estimates had suggested he might have reached six foot but, in a way, this new prediction makes better sense, and means there would have been less of a profound leap in stature between the earlier *Homo habilis* (who were barely any taller than australopithecines) and *Homo erectus*.

Nevertheless, he cuts a tall figure compared with earlier hominins, and he has *long legs*. The two things are connected. That seems like a banal comment, but across modern humans and ancient hominins, there's a connection between stature and long legs: taller individuals always seem to have relatively longer legs. In the origin of our own genus, *Homo*, it could just be that natural selection was acting on body mass and relatively long legs came along naturally with bigger bodies. I think if you saw Nariokotome Boy walking in the distance, you wouldn't bat an eyelid. You wouldn't notice his smaller head and different face; you would just recognise his general body plan as something very familiar.

His long legs would have made him an efficient walker, but anthropologist Dan Lieberman has suggested that there are many features of his skeleton which can't solely be explained as adaptations to walking. For example, he had a strong ligament running down the back of his neck; he had a long, narrow waist and low shoulders; he had big gluteus maximus muscles in his buttocks, and large erector spinae muscles in his back, both of which help to stabilise the trunk; in his leg and foot, he had springy tissues – including a chunky Achilles tendon – to store and release energy; he had short toes. Dan explains all these features as being linked to something very particular, something for which you need springy tendons and ligaments, for which you need the shoulders and trunk to be able to rotate above the pelvis to keep balanced, for which you need strong muscles behind the hip to extend the joint and stop the trunk pitching forward: running. Perhaps those long legs are also 'for running', as increased stride length would mean increased speed in endurance running.

I met Dan Lieberman in his lab in Harvard University. He even put me on his treadmill, to demonstrate the importance of gluteus maximus in running. Gluteus maximus doesn't do much during walking, even on a slight incline, but it's very active during running, contracting to counteract the flexion of trunk on stance-side – stopping you from falling flat on your face – as well as acting to decelerate the leg taking the next step as it reaches the end of its swing. It's hard to say exactly when this muscle grew in size, but Dan argues that it was probably important in the evolution of hominin running proficiency.

Dan's theory, which made the front cover of *Nature* in 2004, is that endurance running was of primary importance to our ancestors. We have focused so much on bipedal walking as the archetypal human mode of locomotion, but it seems that running could have been just as important – at least, by the time *Homo erectus* appeared on the scene. There's probably more to the story than this simple idea of walking and running. Walking up hills and clambering over uneven terrain would also use some of that anatomy which Dan has linked to endurance running. But, on the other hand, running may have provided our ancestors with a way of surviving in a changing landscape, perhaps driving a change in body shape and giving us the basic body plan that we recognise as human today.

Most of us lead pretty sedentary lives today, so it may come as a surprise that we're natural-born runners. But for anyone who has done distance running, it may not be as surprising. It's actually something we're very good at. Our bodies are designed to make running easy and efficient, and while we find it difficult out-sprinting other animals, over long distances humans can outrun many animals, even dogs and horses.

Nariokotome Boy's kind evolved as grasslands – the savannah that is so characteristic of vast swathes of Africa today – were spreading out, and so grazing animals were multiplying and diversifying. While forests were contracting, the expansion of that new grassland environment held out opportunities for hominins that could adapt to it. Making the most of such an environment requires the ability to range over large distances, efficiently, and running may also have meant that our ancestors were able to compete effectively for an important source of energy and protein: meat. Whether that meant hunting or scavenging is a difficult question to answer (perhaps, more likely, both), but the stone tools made by *Homo erectus*, almost two million years ago, appear to be perfect for processing meat – for cutting through flesh and sinew. There are several early archaeological sites in Africa that reveal evidence of butchery in the form of cut-marks (made by stone tools) on animal bones. One recently discovered site, on the shore of Lake Victoria in Kenya, has yielded plenty of butchered gazelle bones, which the researchers argue is good evidence for hunting, rather than just scavenging.

A NEW KIND OF NOVELTY

Human adaptation to running has been suggested to be an example of something which is quite a bold new idea in evolution, an idea which suggests that profound changes in the body can occur *before* genetic changes take place.

Our bodies are adaptable within our own lifetimes. We know that from personal experience: if you were to start a new regime at the gym, your body would change. Certain muscles would grow bigger, and you'd notice those, but what you probably wouldn't notice is that your bones would also change. We tend to think of our skeleton as an inert, lifeless scaffold for our living bodies, but bones are living tissue too. Although heavily mineralised, there are cells living in hollows within the hard bone matrix. These cells are in communication with each other and with cells on the surface of a bone. They respond to altered strains on the bone, laying down new bone where more strength is needed, removing it when loads are lifted. (One of the many challenges for astronauts spending time in a weightless environment is that they lose bone mass.)

As an embryo, as a growing child, and even as an adult, the form and function of your body isn't exclusively determined by your DNA. What this means is that the shape of your skeleton, and in fact, of your whole body, is not just a product of your genes, it's also a product of how you *use* your body: a product of your behaviour. Your genes set parameters within which the shape of your body can change.

Examples of quite profound changes to what might be considered to be the 'normal' anatomy in a particular region of the body can occur when anatomy and physiology is somehow disrupted elsewhere. In 1942, a Dutch vet called E. J. Slijper described a strange case of a bipedal goat. This animal had been born with paralysed front legs so it could not walk on all fours. Instead, it managed to move around quite well by hopping on its back legs. When the goat died, Slijper dissected it and found that its anatomy was very strange indeed: its thorax and sternum were an odd shape, and muscles at the top of the hind leg had changed shape and developed new tendons. Examples like this show just how malleable anatomy can be.

Slijper's hopping goat illustrates an important point: whatever it was that caused the loss of use of the goat's front legs, that problem didn't lead directly to the changes in its hind legs. The anatomical changes came about because of the way the goat was *using* its hind legs. The malleability of anatomy may be more limited during adulthood – you're not going to produce anything this dramatic by going to the gym – but the potential for dramatic change is greater if a behavioural adjustment happens at an earlier stage of development.

The American biologist Mary Jane West-Eberhard has suggested that the anatomical changes which have been linked to running in human evolution might have appeared in this way – as part of how the body adapts to its environment (through changes in behaviour) within a lifetime – and could even have become the 'norm' in a whole population within a single generation. It's an important idea, and in some ways quite heretical. It's certainly a challenge to a strictly neo-Darwinist idea of evolution where random genetic mutations provide the stuff of variation, and then natural selection promotes any advantageous mutations. Anatomical changes in response to new behaviour – or phenotypic accommodation – can also produce variation.

This bold idea about links between behaviour, anatomy and evolution has some echoes of an old hypothesis about how evolutionary change occurs. The pre-Darwinian French naturalist Jean-Baptiste Lamarck suggested something similar, early in the nineteenth century. He believed that animals could pass on characteristics which they had acquired during their lifetimes. The classic example is that of a giraffe's neck: a giraffe stretches higher to reach leaves and its neck grows a little; it passes on this advantageous characteristic to its offspring. Although Darwin was open to this idea of 'soft inheritance', it was eclipsed by the mechanism of natural selection which he proposed in, to give it its full name, *On the Origin of Species by Means of Natural Selection*. This mechanism, natural selection, worked on variation *already present* in populations. That variation was down to heritable factors (later identified as genetic mutations), so changes to anatomy that occurred during an individual's lifetime could not be inherited.

But this now seems too hardline. We now know that the environmental influences on genes can be passed on to offspring. Chemical modifications to the protein packaging of DNA – happening during the lifetime of an animal, in response to environmental stimuli – affect the function of genes. And quite surprisingly, perhaps, those modifications *can* be inherited, even though the genes themselves remain unchanged. This phenomenon of inherited epigenetic effects means that there is something to soft inheritance after all.

When new behaviour leads to a change in anatomy, it's possible that some of the anatomical changes themselves *will* be inherited, at least for a few generations, because of epigenetic modifications which *are* inherited, for at least a couple of generations. But phenotypic accommodation is also likely to lead to true evolutionary, genetic change. Compared with genetic mutations which produce quite random variation (most of it inevitably unhelpful), phenotypic accommodation produces variation which is far from random: right from the start, it's already helping an organism to adapt to its environment. Natural selection will be acting on this sort of non-random variation, as well as on the more random variation produced by mutations.

Individuals with new anatomical features, which have developed in response to behaviour, are likely to have an advantage in survival and reproduction, so the *potential* to produce those changes is selected for. This is something that was first suggested by the American psychologist James Mark Baldwin in the late nineteenth century, in relation to the evolution of learning ability. It's called the 'Baldwin effect'.

Going back to our running ancestors, we can understand how this might happen. Imagine a group of ancient humans whose ancestors didn't go in much for running, but who find that running is helpful to survival. These people live in an African savannah, almost two million years ago. They're eating a wider range of foods than their ancestors. They'll dig for roots and tubers and cook them to make them more palatable. They'll also eat meat when they can get it – and that's where the running comes in. Sometimes they'll hunt animals, but most of the time, perhaps, they get meat from carcasses killed by other predators. They're good at spotting vultures circling on the horizon, and they'll run to get to the carcass before other scavengers arrive. They'll try to avoid confronting large predators but they can throw stones to deter them if they turn up.

In this scenario, running could become really important to a group of humans within a generation. Once a few people start doing it, it's likely that others will join in. Children will copy the adults. Without any genetic changes, the bodies of these people could be changing quite considerably, with bones and muscles responding to the new demands being placed on them. Particularly for those who start running in child-hood, their bodies could end up looking quite different from those of their immediate ancestors. For some people, their bodies may be less 'plastic' than others, and running might be an awkward thing to do, but there are others whose genes provide a latent potential for important changes in their bodies which help to make running more comfortable and efficient. Having started to run, then, these people have influenced their own evolutionary future. The genetic potential for running will be selected for. Over generations, the Baldwin effect means that natural selection will now favour genetic mutations which happen to enhance running performance. But how did all this start? Not with a genetic mutation which was picked up by natural selection, but by a change in *behaviour*.

It could even be that phenotypic accommodation is more likely to produce, or at least, instigate, profound changes in a lineage, compared with the more familiar mechanism of natural selection acting on genetic mutations. Phenotypic accommodation might be an important gener-ator of real novelty in evolution. Novelties are very different from the small-scale adaptive changes which occur as natural selection promotes certain variants and eliminates others. As generators of novelty, the immediate response of an animal to environmental change is a powerful thing – and it can affect many individuals at the same time.

It's certainly not that Darwin was wrong, and natural selection isn't important. He wasn't, and it is, it's just that there might be more to it, and Lamarck might have been on to something after all.

It's possible that our ancestors just started to run without waiting for any useful genetic mutations to help them – those changes may well have followed rather than facilitated the change in behaviour. But however well adapted to endurance running our ancestors may have become, they were still not as fast as many other runners, including some rather large African predators. So it probably paid to hold on to a few more 'primitive' features, like legs that are good for running but not so specialised that you can't still climb up a tree when necessary (imagine a Thompson's gazelle trying to climb a tree!). We've got long, mobile arms, too, inherited from ancestors who spent a lot of their time in trees, but those arms also turn out to be exceptionally useful for throwing things – a skill that would have been useful for our ancestors on those African plains, and which we've inherited.

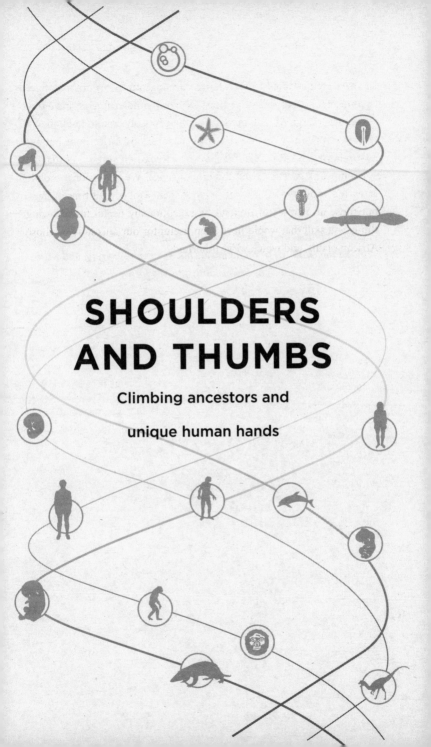

SHOULDERS
AND THUMBS

Climbing ancestors and

unique human hands

'Perhaps nothing is so fraught with significance as the human hand, this oldest tool with which man has dug his way from savagery, and with which he is constantly groping forward'

JANE ADDAMS

THE MOBILITY OF ARMS

Our arms and hands are exceptionally mobile, much more so than the forelimbs of most other mammals. This is something we owe to our tree-living ancestors. The arms of apes are even more mobile than those of monkeys, and this comes down to a different way of moving around in trees. Monkeys are mainly quadrupeds, walking along branches on all fours, with their trunks horizontal. Like most quadrupeds, their chests are flattened side-to-side, with the shoulder blade, or scapula, attached by muscles to the side of the ribcage. Apes move in a different way up in the trees: they climb, hang and sometimes swing from their arms, so their trunks are vertical much of the time, and their chests are flattened front-to-back. The scapula, or shoulder blade, has moved around onto the back of the front-to-back flattened ribcage, so that the shoulders are now positioned right at the sides of the chest. This puts the arms in a position where they can reach out forwards or backwards, above the head or down by the sides of the body. In fact, the shoulder joint doesn't stay in one place; because the scapula is 'floating' on the ribcage, bound to it by muscles, and supported in the front by the strut of the collarbone, or clavicle, it can move to reposition the shoulder joint. You've felt your scapula moving before, but now test out just how much it can move: you can draw your shoulders back further by pulling the scapulae together. You can move the whole shoulder forwards too. And when you move an arm to reach above your head, the scapula rotates so that the shoulder is now directed upwards as well as to the side.

The shoulder joint is in fact the most mobile joint in the body. This makes a lot of sense for an ape, climbing around in trees, because it

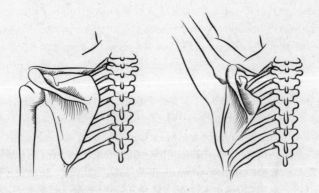

The scapula rotates on the chest as the humerus is raised

enables the arm to reach out in any direction to grab branches. In any joint, there's always a trade-off between mobility and stability. The most stable joint would be one where you had actually fused the bones together so that no movement is possible. The more movement you allow, the less stable a joint becomes. The bony configuration of a mobile joint has to be loose, so you then rely on soft tissues to maintain the integrity of the joint. If you look at the scapula and the head of the humerus (the top end of your upper arm bone), there's actually very little contact between these bones. The shoulder 'socket' on the outer angle of the scapula is very small and shallow: it has about a quarter of the surface area of the joint surface of the humeral head with which it articulates. Ligaments help to support the joint, certainly, but it's muscles that play the most important role in stabilising the shoulder joint. A set of muscles is attached to the blade of the scapula then converges on the neck of the humerus; as well as stabilising the joint, they can also act individually to rotate the humerus in various directions; they're known as the rotator cuff muscles. Despite the best efforts of those muscles, the shoulder is still very mobile and, therefore, unstable by its very nature: it's the most commonly dislocated joint in the body.

Below the shoulder, the elbow is a simple hinge joint, but your fore-arms have extra mobility: your radius, the outer bone in your forearm, can move along its axis so that its lower end spins around to lie on the *inside* of the other forearm bone: the ulna, in a movement called prona-tion. This is a movement which is impossible for many mammals. Most mammals are quadrupeds, and their forearms are usually held in the pronated position, putting their front feet in the right position to make contact with the ground (or tree branches). Think of something like a horse, cow or sheep, these animals have their forearms *fixed* in prona-tion. Some others can half-supinate their paws around – dogs can do this a bit, cats perhaps a bit more – to enable them to grip objects between their paws. But you've never seen a cat turn its paw completely palm-upwards, something which you can do so easily. Again, this mobility is something you've inherited from an arboreal ancestor – being able to twist your forearm to rotate your hand through 180 degrees meant that your ancestors could reach out and grip branches at any orientation.

At the ends of your upper limbs, your hands are incredibly mobile and designed for grasping. All of this mobility allows you to do things you probably just take for granted – reaching up behind your back to scratch an itch between your shoulder blades, swimming breaststroke or front crawl (instead of just 'doggy paddle'), operating a screwdriver, holding a ball or a bowl of soup on a supinated hand – but these would be beyond the reach of most other mammals. You have your ape heritage to thank for this general mobility in your upper limb – your shoulder, arm, elbow, forearm, wrist and hand – but there are also important differences between us and other apes.

HOMININ SHOULDERS

Look at any other ape and they seem to have their shoulders in a perpet-ually 'shrugged' position, almost drawn up to their ears. Inside the shoulder, the joint itself is also 'cranially oriented', in other words it points upwards, which is perfect for reaching up into overhead positions. Our ape cousins are generally much more at home in the trees than we are, with most of them climbing and hanging from their arms, and, if they're gibbons, arm-swinging, on a regular basis.

In contrast, our shoulders have dropped down, and in a resting position our clavicles (collarbones) now lie almost horizontally, rather than steeply sloped upwards and outwards, while our shoulder joints are oriented to the side, not upwards. It's tempting to think that our shoulders dropped down, almost automatically, when our ancestors stopped using their arms so much in overhead positions, but that's surely lazy evolutionary thinking. Why would shoulders just 'drop down'? We need to look for something that could have been the impetus behind a shift to low shoulders. How might wide, low shoulders have provided an evolutionary advantage for our forebears? And when did this change happen?

Instead of a gradual transformation from shoulders like those of other apes to our low-hung human variety, it seems that hominin shoulders were quite diverse, although admittedly the fossil record for this bit of the skeleton is very fragmentary. But it looks like really modern-human-like shoulders appeared on the scene very late in the day.

There are just a few fossils of shoulder blades or scapulae from australopithecine species – those early hominins dating to between four and two million years ago, including Lucy and her kind. These suggest that the shoulder socket was oriented upwards, much more like that of other apes than yours or mine. Other features suggest that these ancient scapulae were positioned high up on the thorax. By the time Nariokotome Boy strides on to the scene with his long legs, one and a half million years ago, his shoulder joint no longer faces upwards, but it doesn't face straight out to the side either. Nariokotome Boy had very short clavicles – more like those other apes. Nariokotome Boy was, after all, a boy, but it's predicted that even if he had reached adulthood his clavicles would still have been relatively short compared with ours. So, while his shoulder joint might have dropped down on the thorax to lie at a level similar to yours or mine, his shoulders didn't sit right out to the sides but instead lay in a more anterior or 'forwards' position.

It's been argued that low, wide shoulders could have developed because they provided an advantage in running, allowing effective counter-rotation of the upper body against the destabilising, twisting force of one foot hitting the ground, then the other. We've seen other potential adaptations to running in Nariokotome Boy's skeleton, so perhaps his shoulders are part of that. But that counterbalance wouldn't have worked all that well in Nariokotome Boy, as his short clavicles suggest that his shoulders would have still been quite narrow. This would also have limited his ability to stabilise his head while running.

In later hominins, including *Homo antecessor* (a species known from fossils from the Sierra de Atapuerca, in Spain, dating to about 800,000 years ago) and Neanderthals, clavicles are long, even longer, in fact, than in modern humans. Long clavicles push the shoulders back into a position where the shoulder joint sits laterally, at the side of the chest.

It's been suggested that this change in shoulders came about because of advantages linked to a very specific use of the shoulder – in throwing, and this does seem to be something that humans are particularly good at.

This isn't to say that other apes don't occasionally throw things, some-times even fairly accurately. I remember filming a *Horizon* programme at the wonderfully named Pongoland – the ape enclosure within the Primate Research Center at Leipzig Zoo. One of the orangutans took a dislike to our director. There's no glass between you and the apes at Pongoland, you walk through the enclosure on a raised walkway with a low wall. The director and our sound recordist ended up on the receiving end of several, very well-aimed, handfuls of excrement.

Nevertheless, there's a combination of anatomical features in your body which are missing in other apes, and which suggest that overarm throwing *must* have been important to our ancestors, and has shaped your anatomy. These features include laterally positioned shoulders, and it's easy to imagine how shoulders positioned right out at the sides of the ribcage would have provided an advantage in throwing, enabling the whole arm to be moved far back, ready to throw. But together with other features – a tall, flexible waist which allows rotation of the torso, and a twist along the axis of the humerus so that the elbow is angled slightly outwards – that laterally facing shoulder joint provides another advantage in throwing. It has been shown experimentally that these features all help to stretch tendons, ligaments and muscles around the shoulder when the arm is pulled back ready for an overarm throw. The act of throwing uses many muscles, activated one after another in a chain, starting in the hips and trunk, flowing up into the shoulder, elbow and wrist. If you go through the motions of performing an overarm throw, you can feel just how many joints are involved. Now slow it down. You start by standing, one foot in front of the other, and bring your arm back into position, ready to throw. Then go: you start moving from the hips; your pelvis is firmly connected to your spine at the sacroiliac joint so that the twisting movement now continues all the way up your spine. Small movements between each vertebra and the next add up to a lot of rotation by the time you get up level with your shoulders – they've now twisted through 90 degrees or more. Now your arm is moving forward, in front of the shoulder, using the incredible mobility in that ball-and-socket joint, and your spine is still twisting, pushing that shoulder even further forwards. Now the hinge joint of your elbow is straightening and, finally, your flexed fingers uncurl to release the imaginary ball or spear from your grasp. Do this quickly (maybe outside) with a real ball, or if you're feeling brave, a real spear, and the elastic energy stored in the stretched tendons and ligaments around those joints is released as the arm moves overhead, helping to boost the power of the throw.

It's very possible that throwing is what gave wide shoulders to *Homo antecessor*, *Homo heidelbergensis*, Neanderthals and to *you*. (And, conversely, it means that Nariokotome Boy, like modern people with short clavicle syndrome, would not have been able to throw well.)

Now it's all very well saying that earlier hominins like *Homo antecessor could* have thrown – given their wide, laterally sited shoulders I think it's fairly convincing, because it's difficult to think of reasons why shoulder shape and position should have changed so dramatically in the way they did if you discount throwing. But there's no convincing some people – they want to see direct evidence. That's hard to come by for an activity. Footprints are pretty good records of walking on two legs, but you don't get any impressions of *throwing* in the same way. If our ancestors were just picking up stones or sticks and throwing them at prey, or at other predators to drive them away from carcasses, there would be no record of this activity in the archaeological record. The thrown stones would look like any other stones and the sticks would rot away. But when humans start to make objects that were specifically *designed* for throwing – *spears* – then we can infer that this is what they were actually doing with their laterally placed shoulders.

The invention of projectile weapons must have been such an important advance. Imagine the difference it would have made to our hunter-gatherer ancestors. Spears had probably been around a while, used as thrusting weapons, to kill prey or defend against predators, but moving on to throwing these weapons would have made them more effective, while helping to keep the hunters out of harm's way.

The problem is, it's really difficult to prove that a spear was designed for throwing! Some early potential evidence is a collection of wooden spears from Schöningen, in Germany, dating to around 400,000 years ago. These spears are apparently weighted for throwing, and seem to be fairly convincing, if indirect, evidence for humans being able to throw by that time, at least. But some researchers still doubt that the Schöningen spears really were used for throwing. Very recently, some other evidence has come to light. It's a bit younger, but more conclusive.

Archaeologists have pored over ancient stone points to try to work out whether they might have been used as spear tips, and not only that, but as tips on spears which would have been thrown. The interpretations have been largely based on ethnographic comparisons, looking at similar tools made by modern hunter-gatherers, and on the physical properties of the stone points – their size, shape and weight. But archaeologists have recently started to apply ballistic science to the problem, looking for telltale fracture patterns which reveal how fast the point was travelling when it hit its target. The latest evidence comes from the Rift Valley, in Ethiopia, where archaeologists discovered thousands of fragments and points, made from a natural, volcanic glass called obsidian, at a site called Gademotta. The fracture patterns on the obsidian points, caused by impact, indicate fracture velocities of 1000 metres per second, which means they were probably javelin tips.

The obsidian points date to almost 300,000 years ago, which shows that projectile weapons were being used by at least this time. Hafting stone tips to spears or javelins and using them as projectiles has also been thought of as a uniquely modern human thing to do. Strictly speaking, modern humans (that's not just any old *Homo*, but our very own species: *Homo sapiens)* didn't appear on the scene until after 200,000 years ago, but the Gademotta points are telling us that this aspect of 'modern' behaviour had developed before 'modern' anatomy. This is important: we shouldn't expect behavioural modernity and anatomical modernity to suddenly appear overnight, to arrive as a package. The features that characterise our species today would have appeared piece by piece, in a mosaic fashion.

APE ELBOWS, WRISTS AND HANDS

Your elbow is broadly similar to that of a living ape. Ape elbows are different from monkey elbows; in both, the surfaces of the joint at the end of the humerus are separated into two parts: a ball-shaped capitulum (or 'little head', in Latin) and a spool-shaped trochlea ('pulley' in Greek), articulating with the radius and the ulna respectively. The ball of the capitulum forms a joint with the dish-shaped head of the radius, allowing it to move in flexion and extension alongside the ulna, but also enabling the radius to spin along its axis to produce the movements of supination and pronation.

These movements of the forearm – supination and pronation – are, to me, some of the most incredible things we can do with our bodies. Most mammals *can't* do them, but it's something you take for granted. We owe this ability to our tree-living primate relatives, but today, we use this movement to help us manipulate all sorts of things. Twisting our forearms allows us to place our hands so that the palm faces in different directions. Try it out for yourself: sit down, with your elbows tucked in to your waist, and your forearms and hands lying parallel with your thighs. Bring your hands to lie palms-up, your thumbs out to the sides. Your forearm is now in a supinated position. Now rotate your hands inwards, so that your hand lies palm-down on your thigh. You've just pronated. Do it again, and this time, look carefully at your forearms and wrists. The rotation isn't happening at your wrist joint; it's happening in your forearm. Your radius starts off lying on the outer side of your forearm. When your forearms are pronated, palms down, the distal or wrist-end end of your radius has twisted *over* the ulna so that it now lies on the inside. In other words, your two forearm bones, having lain parallel, side by side in supination, are now crossed over in pronation. This special movement allowed your tree-living ancestors to reach out and grasp branches at almost any angle. Now, you wouldn't be able to use a screwdriver without it.

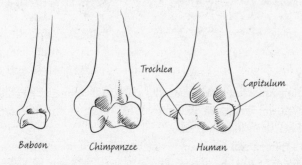

The lower end of the humerus in a baboon, chimpanzee and human

Apes use their arms in different ways from their monkey forebears, and they need good, strong elbows. At the end of the humerus, the shape of the trochlea in apes (including us) is quite distinct from that in monkeys. The spool-shape is pronounced, with steep edges which help to keep the ulna very stable through a wide range of motion – very useful for a primate which is using its arms for climbing and hanging from. Apes also have a deep pit in the back of the humerus, accommodating the olecranon of the ulna (this is the knobbly bit at the back of your elbow) when the elbow is fully extended.

But as we get to the end of the limb, to the wrist and hand itself, there's a big difference. Here's what the bones look like in your hand, alongside those of a chimpanzee:

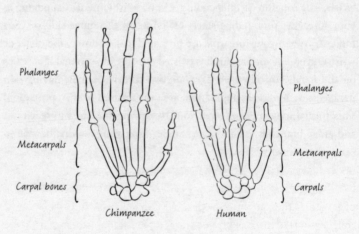

Hand bones

So chimpanzee feet look like hands (to us), and chimpanzee hands look like, well, not very much like hands. What's that tiny thumb about? And those massively long fingers?

At a chimpanzee sanctuary in Entebbe, on the shores of Lake Victoria, in Uganda, I had the opportunity to look at chimpanzee hands very closely. With the film crew, I was allowed onto the chimpanzees' island through a tall, chimp-proof gate, then the chimpanzees were let in to join us. I crouched down to meet them; I had pocketfuls of peanuts and they realised this immediately – I was their new best friend. They were very used to humans and very used to being handled, so they threw themselves onto me – the littlest one, a three-and-a-half-year-old called Nipper (who lived up to her name!) climbed all over me, and when I stood up, she clambered up my legs and onto my back, wrapping her long arms around my shoulders. We went down to an artificial 'termite mound' the keepers had built – a metre-high dome of mud with plastic tubes containing some honey. The chimpanzees knew this game and started picking up bamboo sticks to thrust into the holes and retrieve the delicious treat.

It was fascinating looking at how they used their hands, holding the sticks in a grip with all their fingers curled around, rather than trying to hold the stick between fingers and thumb as we might do. Pick up a pencil and you'll immediately understand the difference.

I tried to do a 'piece to camera' by the mound, while Nipper jumped on my back and bit my arms (quite hard, but playfully and not menacingly). Whereas the chimpanzees in the forest at Kibale, where we had also filmed, practically ignored humans, these chimps – who had been rescued from the pet trade or from traditional doctors, or found badly injured in the forest – seemed to see humans as something very much like them. At least, the youngsters certainly saw us as things they could play with as they might do with other chimpanzees. It could get quite rough, as one of the keepers had warned me, and the oldest of the three chimpanzees we were filming with, a six-year-old male, was pretty strong. He ran up and whacked me quite hard a couple of times. I was lucky not to lose my radio-mic; the cell-phone-sized transmitter was tucked away in a waistband under my vest top and shirt, but the chimps knew there was something there and kept reaching up the back of my shirt to try to grab it.

I sat down with the chimpanzees to feed them some fruit: avocados, which were cast aside, and oranges, which were seized, broken open and devoured. I had been reading papers by Mary Marzke and Matthew Tocheri about human and chimpanzee hands. Handling food was an activity which allowed researchers to watch how chimpanzees would manipulate objects, and compare the range of grips with those used by humans making stone tools. The same range of grips had been observed, although chimpanzees seemed to have less powerful thumbs, whereas humans could combine precision with strength when handling objects. Marzke noted that the 'pad-to-side' or 'key' grip, with the thumb pushing down against flexed fingers, was important in human tool-making and tool-use. Chimpanzees used this grip too, but often supported a piece of food with the other hand or a foot as well, rather than holding the object firmly with a single hand. Similarly, wider grips used to hold spherical objects seemed less strong against resistance in chimpanzees compared with humans. Whereas we will happily hold an apple in one hand and take a bite from it, chimpanzees might use another hand to help pull the fruit away from the closed teeth.

Having read that research, it was fantastic to sit and watch the older female chimpanzee, Sarah, as she used both hands to hold an orange, between the thumb and fingers, and bite at the juicy segments inside. Her thumbs were very short, only just reaching up to the knuckles (metacarpophalangeal joints) of the fingers. Those thumbs were also slender compared with the long, robust fingers. Although we had the same basic anatomy – the same bones and muscles – in our hands, there were obvious differences between us. It's tempting to say that those differences relate to Sarah's predisposition to knuckle-walking, where she might need those robust fingers to bear the load, whereas in my hand, a more robust thumb (and little finger, across the other side of the hand) might have arisen as an adaptation to tool-making and tool-use. These are reasonable suggestions, but if we want to pin these down as adaptations, then we need to know exactly how these particular anatomical features relate to function. We also need to know how our respective hands have changed since our last common ancestor.

THE HANDS OF OUR ANCESTORS

Once again, Ardi's fossil bones, from 4.4 million years ago, are helpful when it comes to trying to understand what the hand of the last common ancestor of chimpanzees and humans might have looked like. And once again, we can reject the idea that the ancestral hand looked like a modern chimpanzee's hand. Their hands have changed as much, if not more, in the millions of years since we last shared a common ancestor.

Ardi's fingers were long and curved, but not as long as chimpanzee fingers. Long, curved fingers are characteristic of gorillas and chimpanzees, and seem to be related to grasping branches and hanging from them. Juvenile chimpanzees and gorillas have very curved phalanges (finger bones), and their fingers grow straighter as they become adults. This reflects differences in the way juveniles and adults move around. Young gorillas spend much more time up in trees compared with bigger, heavier adults. It also suggests that curved fingers may develop *in response* to the way bones are being used. Curvature of finger bones seems to be a genuinely useful adaptation for apes hanging around in trees: it helps to reduce the strain on the bone.

In gorillas and chimpanzees, the metacarpals, the five bones projecting to the digits but lying within the hand itself, are long, whereas the 4.4-million-year-old hominin, Ardi, has short metacarpals. This suggests that the metacarpals evolved to become longer independently in gorillas and chimpanzees, probably as an adaptation to vertical climbing and arm-hanging. This is an example of convergent evolution, when a similar anatomical feature appears in separate lineages, usually because these different species are living in similar environments and have similar ways of life. Short metacarpals, such as Ardi's, instead allow a hand to mould around a branch, which is useful in careful climbing. Ardi has a reasonably robust thumb compared with modern African apes. Once again, we're seeing an image of the last common ancestor of humans and chimpanzees which is decidedly un-chimp-like – not a knuckle-walker, not accustomed to climbing vertical trunks or hanging from its arms, and in some ways, perhaps, more like an orangutan than a chimpanzee, but actually unlike any living ape.

Lucy and her kind, the Afar australopithecines, who lived more recently than Ardi, between four and three million years ago, still had curved finger bones. Although there seems no doubt that these australopithecines were fully bipedal, their curved phalanges suggest that they were also spending at least some of their time in the trees. Perhaps Lucy still climbed trees to reach particular fruits, and made a tree-top nest to sleep in each night, like chimpanzees. Don Johanson, the discoverer of Lucy, made such suggestions but also warned that Lucy's curved fingers could just be evolutionary baggage. Other elements of Lucy's anatomy suggest that this could indeed have been the case. The anatomy of her spine perhaps suggests that she was unlikely to have been spending much time moving around in trees. In chimpanzees and gorillas, a short, stiff spine reduces the risk of injury and helps with moving between trees. In contrast, Lucy had a long, flexible lumbar spine, which suggests that she had virtually abandoned life in the trees.

Looking again at hands, Lucy's fingers were shorter than those of African apes, and a little more like ours, and her thumb was broad at the tip but still fairly weedy compared with our thumb. However, at least one australopithecine species had more 'human-like' hands. Hand bones are generally quite few and far between in the fossil record, but an almost complete wrist and hand skeleton was found at a site in South Africa, dating to almost two million years ago, and attributed to the species *Australopithecus sediba*. This hand looks strikingly 'modern', with short fingers and a long, robust thumb. It's tempting to think that this australopithecine must have been doing something very special with its hands. Although stone-tool-making has long been considered to be the hallmark of our genus *Homo* (so much so, that one of the earliest members of this group is named *Homo habilis* – the 'handy-man'), early dates for stone tools mean that we have to accept that some australopithecines were making tools. When it comes to later hominins (like *Homo antecessor*, from Spain, or Neanderthals), we have masses of evidence of

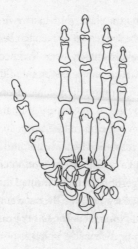

The human hand

the tools they made, and their hands are even more like ours (though very robust), with short, straight fingers and strong thumbs.

UNIQUE THUMBS

The difference in human thumbs compared with other living primates goes beyond bony anatomy. You and I have certain muscles in our thumbs which are lacking or different in other primates, even in our closest living relatives, chimpanzees. You have a muscle tucked in between the first and second metacarpals which powerfully draws the thumb in towards the other fingers. This 'adductor pollicis' muscle becomes very developed if you're accustomed to applying force with a tool that you grip in your hand. I've noticed that archaeologists develop large first palmar interosseous muscles when they're on site and moving considerable volumes of earth with a hand trowel. It's something that really stands out when you place your hand palm down on a surface; this interosseous muscle bulges up in the cleft between the thumb and the index finger. I have honestly seen archaeologists competitively comparing the size of their interosseous muscles in the pub. In fact, adductor pollicis is not a uniquely human muscle, but there's a small muscle which sits alongside it, assisting it, that is. It is, rather snappily, called the first volar interosseous of Henle.

There are two more muscles which are unique to human hands: one is an (additional) extensor which pulls the thumb out to the side, and the other is a muscle which originates in the front of the forearm, with its long tendon travelling across the wrist, into the hand and inserting into the last phalanx of the thumb: a powerful flexor. Other primates have this muscle, but it's part of the deep flexor muscle which also sends tendons to all the fingers. In humans, flexor pollicis longus is a separate muscle in its own right. It's a rather beautiful muscle, too; the tendon starts in the forearm and the muscle fibres attach to it at an angle so it looks like a quill pen.

Why is flexor pollicis longus singled out for special treatment in humans? What is it doing? The basic answer, of course, it that it flexes

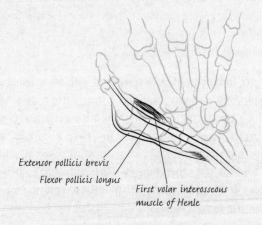

Extensor pollicis brevis
Flexor pollicis longus
First volar interosseous
muscle of Henle

Uniquely human thumb muscles

the thumb. Many researchers have suggested that this flexor is crucial for one particularly human activity: tool-making.

Our consummate command of technology is unique. Other animals produce material culture, so it's not an absolute difference between us and other species, but having said that, the degree of difference is such that it really does set us apart. The technology we have today seems a world away from the earliest evidence of technology, in the form of stone tools, but another hallmark of humans lies in passing on knowledge and skills – we build on the culture we inherit. The earliest stone tools come from Ethiopia, and date to 2.6 million years ago. So it seems obvious that certain peculiarities of human hand anatomy might reflect that ancient tradition of tool-making. Many researchers have made a link between tool-making and our unique flexor pollicis longus muscle, and presumably a *robust* thumb would also have been an advantage to a tool-maker, as it would be good at resisting the high forces generated by striking one stone against another to create flakes.

I had heard about some interesting new research looking into the structure and function of the human hand, and so I travelled to the States, to George Washington University, to talk to Brian Richmond and Erin Williams. They were investigating the stresses experienced in the hand during stone tool-use and tool-making. In particular, they wanted to test the hypothesis that unique aspects of human thumb anatomy were related to tool-making.

When I arrived in their lab, they took the opportunity to turn me into another test subject. Erin fixed thin, tape-like pressure transducers to my fingers and thumb. Once fully wired up, I tried making a crude stone tool while Erin recorded the pressures experienced in my hand. Then I had a go at using the stone flakes I'd made to cut meat. I probably wasn't the best subject to show pressures in the hand of a stone tool-maker – it wasn't something I'd ever spent much time doing, and it's harder than you might imagine to smash one stone against another and get what you want in terms of a stone flake. Thankfully, there are people around today who have spent time developing the knapping skills that we know (from their tools) were possessed by our palaeolithic ancestors. Brian and Erin had already had a chance to study pressures in the hands of accomplished knappers, and their findings were unexpected: pressures were high in the fingers, especially in the index and middle finger, but not particularly high in the thumb.

Instead – and my results reflected the general findings of this research – stresses on my thumb shot up when I started to use a stone flake. I naturally held the flake against the side of my curled-up index finger, with my thumb pressing down to hold it in place. The pressure measurements showed that I was having to apply significant pressure with my thumb to hold the flake in place as I used its sharp edge to cut down through a joint of meat. *This* was when the pressure in my thumb was higher than in my other digits, not during tool manufacture.

This is interesting, as compared with chimpanzee hands (and presumably those of the last common ancestor) our thumbs are much more robust and powerful in pinching grips. Part of that pinching power comes down to our unique flexor pollicis longus muscle. It seems that our strong thumbs may have evolved through selection pressure related to *using*, rather than making, stone tools.

Look at your hand. You can see echoes of very distant ancestors, as well as features that you share with other primates, and characteristics that seem to be special to us. You have five fingers, a primitive number of digits, inherited from an ancient ancestor after all that experimentation with fingers and toes by creatures like the early, newt-like tetrapod, *Acanthostega*. The basic musculature allowing you to move your wrist and your fingers is also shared with living amphibians. Unlike those creatures, though, you have opposable thumbs. This isn't a uniquely human characteristic, but something shared with most primates. You also have fingernails rather than claws – another primate feature. But then there are also all those more recently evolved characteristics, including a robust thumb and a unique muscle operating it, which have sculpted your hands into organs capable *of* sculpting, capable of manufacturing and wielding tools. As you look at it, and move it, your hand has all that evolutionary history contained within it.

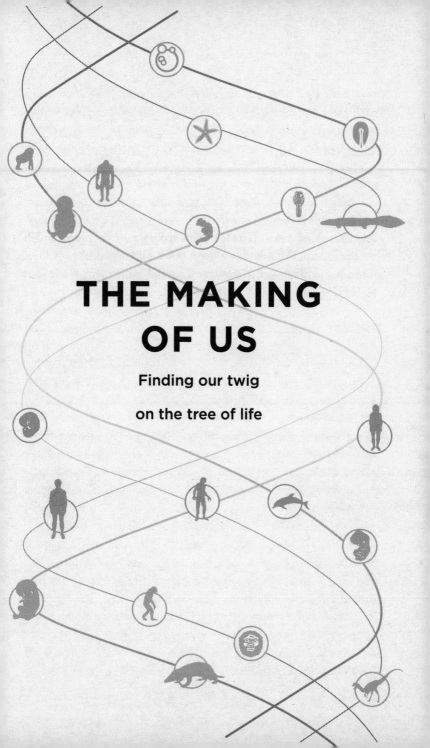

THE MAKING
OF US

Finding our twig

on the tree of life

'*'Tis the sublime of Man,
Our noontide majesty, to know ourselves
Parts and proportions of a wondrous whole'*
COLERIDGE

At the end of this anatomical journey you can look at your hand and see not only something which developed out of a minute limb bud in your own developing embryo, but something which evolved from a fish's fin, over millions of years and millions of generations. However different you and I may feel ourselves to be from other animals, each of us has been shaped by the same forces acting on the rest of life on this planet. You're the product of the developmental mechanics that created an entire human body out of a single, fertilised egg. On a grander time scale, you're a product of evolution; from your colour-sensing eyes to your vocalising larynges and your smiling face, from your hearts and lungs, down to the ends of your short, lined-up toes and out to the tips of your robust thumbs and dextrous fingers. There's a deep history hidden in your anatomy. As Terry Pratchett so wonderfully put it: 'We *are* history.'

Having my own children reawakened that sense of wonder in me: how astounding to have played host to that extraordinary act of creation, to have had that incredible origami-like process unfolding inside me, using resources from my body to build a tiny new one, until, after nine months, a baby emerged into the world. And we're still trying to understand how that happens. It's a fundamental question, of course, and one that all children ask, at some point: 'How are babies made?' You are then obliged to formulate a response, trying to be honest while attempting to make the thing seem both credible and vaguely palatable for a child.

We've already discovered many secrets of our own recent histories – the story of our embryological development. We've known for some time now how an embryo springs into being, from the union of one egg and one sperm; then a ball of cells transforms into a disc, curls up into a cylinder, sprouts limb buds, while cells inside the tiny growing body are proliferating, migrating and dying to sculpt the tissues and organs of a human body. But it's only much more recently that embryologists have been able to understand *how* this process happens: what's driving cells to divide, move around or commit suicide. We're now starting to develop an understanding of how DNA directs development. Geneticists have only recently been able to sequence a whole person's genome, and understanding how all the genes work together and interact with the environment to transform an egg into a baby (and eventually, an adult) is still very much a work in progress. The more we discover, the more complicated it seems to become, so that it's not just the genes themselves that are important, but the stretches of DNA in between, which play a part in regulating the genes. On top of that, DNA can become chemically modified during our lifetimes, producing another level of regulation and another layer of complexity which demands its own field of study, in epigenetics.

It feels as though we've come such a long way since Aristotle, but there's still a long way to go until we really understand this biological miracle. And although the old ideas of epigenesis and preformation both seem out of date in the light of what we now know about embryological development, it turns out that both contained a germ of truth.

When microscopy allowed biologists to *see* sperm and eggs, it became clear that the beginnings of a life involved much more than the simple mixing of fluids, as Aristotle's theory of epigenesis had suggested. But then neither the sperm nor the egg appeared to contain anything which could be interpreted as a miniature human inside it. A complete preformed human existed neither in the gametes, nor in the fertilised egg. An early embryo looked nothing like a baby, let alone a miniature adult. In the eighteenth century, Caspar Wolff saw complexity developing from simple beginnings, as a ball of cells formed and then gradually transformed itself into differentiated tissues. Detailed structure appeared gradually in the embryo, apparently unfurling naturally like ice ferns appearing as water freezes on a pane of glass.

Wolff was right – embryogenesis is all about differentiation, starting with stem cells and ending with differentiated tissues – forming skin, muscle, bone, cartilage, nerves and the rest. But the first cell, the fertilised egg, is not as simple as it appears under a light microscope. We now know that it contains a code – the DNA needed to build a whole body. So in some ways, are we turning full circle, coming back to preformationism again? There may not be a tiny preformed human inside the egg, but there's certainly complexity – there's a recipe for a human in there. The 'idea' of a human body is already there, right from the start, written out in the four letters of the DNA code. This 'idea' isn't a blueprint or a plan for a body, though, it's a series of instructions which will lead to the generation of form. In this way, it really is like origami: a set of quite simple instructions can lead to the production of a complex shape.

So this is where we're at then: epigenesis happens, with simple things turning into more complex ones, but not like ice ferns, it's much more directed: the DNA in the fertilised egg takes control and provides all the instructions the embryo needs to know in order to grow in complexity. But this is too simple and too neat an explanation: it suggests that genetics are everything when it comes to development, and we know that's not true either. Genetically identical twins don't develop in *exactly* the same way, and the differences arise because of the interplay between DNA and the environment of a living being. The environment plays a role during intrauterine development but becomes even more important when the baby is born. Your DNA doesn't minutely dictate exactly how your body (including your brain) will develop, it provides instructions which have built-in flexibility: it sets parameters. This is important, because it means that innovations can occur, sometimes involving fairly major changes to a body, without any genetic mutation. Think of that hypothetical example of the human body changing as our ancestors took up running.

Our embryological development is deeply, inextricably intertwined with a much older history. You may not have run through the sequence of your entire evolutionary history during your embryonic development, but it's impossible to ignore those embryonic echoes of ancient ancestors. There were fleeting impressions – of larva-like segments, gills, fish hearts, the lancelet brain – as you underwent that incredible transformation from a single egg to a complex human body. Evolutionary developmental biology (or 'Evo-Devo') is all about bringing together evolution and embryonic development: phylogeny and ontogeny. The parallels between these two developmental histories – changes in species over time and changes during the embryonic development of a single individual – were recognised long ago, but advances in molecular biology mean that biologists are now uncovering the genetic basis of changes that are happening in evolution and embryonic development.

While Ernst Haeckel proposed that embryos recapitulated their evolutionary history, Karl Ernst von Baer's idea of divergence and differentiation was different: embryos contained echoes of the *embryos* of earlier ancestors, not of ancestral adults. This makes sense if we think of evolutionary change as involving the addition of characteristics, or a change in timing within development.

Archaeologists often speak of a site being a palimpsest of different phases, superimposed on each other. It's a lovely analogy: a palimpsest is an ancient manuscript which has been erased and written over: 'palimpsest' means 'scraped again' in Greek. For an archaeological site, the analogy of an ancient manuscript, scrubbed clean(ish) and written over, with each generation or culture leaving its mark, works well. Embryological development is like a much more modern palimpsest; the analogy works if we think about a document written in word-processing software – the document changes over time, new words can be added or deleted anywhere in the text, and these small changes might significantly alter meaning. It's not surprising that, as you read the document, you'll be struck by similarities to earlier drafts. An early human embryo – that ball of cells which forms in the first week of development – looks pretty

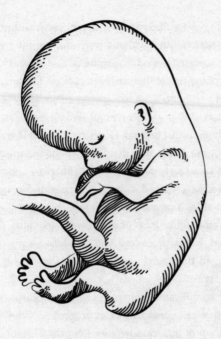

An eight-week-old human embryo

much like any other vertebrate embryo. The early heart of a human embryo looks very much like an embryonic (and adult's) shark heart. But by eight weeks after conception, you looked recognisably human.

THE UNLIKELINESS OF BEING

Evolution proceeds by tinkering with the embryological development of organisms. That sounds like a dangerous thing to do – and it often is, for many genetic mutations will produce completely unviable offspring. This probably happens much more often than you might think. In fact, it's impossible to know how many human pregnancies are unsuccessful, as abnormal embryos are likely to be lost very early, within two or three weeks of fertilisation, often without even affecting the menstrual cycle and leading to a missed period. It's estimated that around *half* of conceptions are lost, many before they are even noticed. Of those unsuccessful pregnancies, it's estimated that half involve major chromosomal abnormalities. Miscarriages work as a natural screening programme; without this loss, it's thought that around 12 per cent, rather than 2 per cent, of babies would be born with birth defects. But of course miscarriages don't just happen in those early days and weeks either: around a fifth of pregnancies detected by a test end with a miscarriage, before twelve weeks.

It's not something that parents often talk about but it does illustrate again the extreme unlikeliness of your own being; after *that* particular sperm fertilised *that* particular egg, you found your way to your mother's uterus, implanted there safely, and carried on developing and growing until you were born.

Given how complicated embryological development is, involving a conversation between genes as well as properties of developing tissues which emerge out of how closely or loosely cells stick together, it seems perhaps quite remarkable that genetic mutations don't derail it more often. It's important to realise that there are these limits to tinkering. The possibilities for changing an already complex organism are quite tightly constrained. The constraints inherent in developmental mechanics, in what works and what doesn't work from a gene's eye view, as well as the constraints imposed by the nature of biological materials, mean that the future for any species isn't full of limitless possibilities. Once again, natural selection doesn't seem to be quite as 'in charge' of the direction of evolution as we might have imagined. It has a limited selection to select from.

The constrained nature of embryonic development also means that individual features are not free to vary, independent of each other. We have to be careful when anatomising a body and looking at each of its features from an evolutionary perspective: individual features didn't appear in isolation, they were never designed to work on their own, and something which looks as though it could be an important anatomical feature, because it's different from what we see in other animals, could actually just be a quirk of development.

Stephen Jay Gould used an architectural analogy to illustrate this point, with the example of the Basilica of San Marco in Venice, where there are approximately triangular spaces between the arches and the dome above. These architectural features are called 'spandrels'. They're not an important part of the design that the architect aimed to include, they just happen to be there because there's inevitably a gap between the edge of the arches and the rim of the dome. But once they're there, they're made use of. In the example of San Marco, they're used for depictions of the four evangelists.

Gould's point was that not all features of organisms have arisen directly through natural selection. Many features 'come along for the ride' – they're side effects, like spandrels, of other things which *have* been selected for (those things being the biological equivalents of the arches and dome of San Marco). It's important not to be hyper-adaptationist, because not all anatomical and behavioural traits have been selected for. There will be elements of the 'design' of your body which really are spandrels – perhaps they need to be there for structural reasons, rather like the spandrels of San Marco, or perhaps they're there because of the limitations of the biological materials used to make your body, or the limited genetic information used to build it. These constraints have an important role to play in determining the shape of your body and its parts. Natural selection can only select among possible variants, and variation is already limited by constraints, before natural selection gets a chance. We have to keep an open mind about the possibility that some features of our bodies may be spandrels rather than adaptations. Even some of the characteristics which Dan Lieberman pointed to as specific adaptations to running *could* just be spandrels. Long legs, bigger bum and back muscles might just be side effects of an increase in body size in *Homo erectus*. As we learn more about the genetics of development, we should be able to work out if a feature is more likely to be an adaptation, rather than a spandrel.

The evolutionary history of an organism and the mechanics of the developing embryo will both impose constraints on development, so that, at any point, evolution has a limited set of paths to take. Together with the likelihood of closely related animals living in similar habitats, this means that we shouldn't perhaps be so surprised by examples of convergent evolution, when we find an evolutionary novelty springing up, quite independently, in several different lineages.

As a quite astonishing example, it looks like the mammalian middle ear – with its three ossicles – evolved at least *four times*, in four separate but closely related lineages. And, going back to our ape cousins, perhaps it's not at all surprising that similar ways of moving around seem to have sprung up independently in separate but closely related lineages. For some large-bodied primates moving around on the ground, knuckle-walking was an obvious solution. Up in the trees, hanging from arms, as well as weight-bearing on two legs, made sense. Adaptations to these behaviours probably arose in parallel, rather than being inherited from a common ancestor. Down on the ground, it seems that earlier apes like *Oreopithecus* may also have walked around on two legs, like us hominins. Bipedalism is probably not as special as we've previously thought – in fact, it's very likely that it appeared several times, in different lineages, just like the mammalian middle ear did. The developmental constraints were in place, and the 'zeitgeist' was right – among several closely related species, living in similar environments – so it's not at all surprising that a similar solution might have popped up several times.

It might seem that constraints make evolution very predictable if, once a certain set of features is in place, it's more likely that others will appear. But there are chance events which have a huge impact on the way species evolve, with none of the sieve-like precision of natural selection. Catastrophic mass extinction events can wipe out species, or whole groups of species, which have been very successful up until that point and change the future prospects for other species. The catastrophe that spelled the end for the dinosaurs was just such an unpredictable event. It doesn't matter how evolutionarily fit you are if a meteorite lands on your head; 66 million years ago, the Chicxulub asteroid slammed into the Yucatan peninsula of Mexico, and the dinosaurs' days were numbered. The fallout from such extinction events is also impossible to predict, but in the aftermath of the Chicxulub impact, a small group of surviving animals – our own mammal ancestors – was able to diversify and occupy the ecological niches which had been cleared out by the impact. If Chicxulub hadn't happened, it is vanishingly unlikely that humans would ever have evolved.

MASTERS OF OUR OWN DESTINY

When we look at other animals, we might see creatures that seem to be helplessly swept along by an evolutionary current, adapting to changing environments, and just going with the flow. Then we look at ourselves and see a species which seems to be able to fight the tide: we mould our environment to suit *us* rather than letting ourselves be moulded by our environment. To some extent, this is true, but we're not unique in modifying the environment we live in, and things we do to change our environment can actually end up changing us.

Your hands are quite different from those of a chimpanzee. Your fingers are shorter, your thumb is longer and more robust. The special design of your hand may have appeared as genetic mutations led to improvements on changes which had occurred very quickly in the hands of your ancestors as they started to use stone tools. It's easy to imagine how the shape of human hands might have accommodated (within a lifetime) making and using stone tools. It's also easy to imagine how tool-use might have led to improved chances of both survival and reproduction, so that natural selection would then instigate genetic, evolutionary changes.

But from those first, accommodating changes in muscles and bones, to the genetic changes which enshrined that adaptation in your genome, this is about more than a direct interaction between an organism and its environment. There's something in the middle, and that something is culture, or technology. Our hands have been shaped by the tools that our ancestors made and used. We're so used to thinking of our bodies as adapted to an environment. We think of natural selection effecting the 'survival of the fittest' – with a survival advantage possessed by individuals who best fit their environment. I think that whenever we use that word 'environment' in relation to evolution in this way, we can't help thinking of the natural environment. We're thinking of this animal fitting its environment, which might include certain types of terrain, particular plants, perhaps prey and predators as well. But of course that's not the whole context of our animal. Its environment also contains other animals from its own species, each of which may be a competitor, an ally, an enemy or a mate – or playing all of those roles at different times. Our animal must also be a good 'fit' in its social context in order to survive and thrive.

The idea that animals influence their own environment, outside the body, in a way which is important to their own survival and evolution, is something that Richard Dawkins called the 'extended phenotype'. Of course, many animals also modify their environments in particular ways which help their survival; foxes and rabbits dig holes in the ground to live in, to hide from hunters and predators, to help them conserve heat in cold nights and winters, and to provide somewhere safe to raise relatively naive and helpless offspring. It's easy to imagine that, for a fox or a rabbit, being able to dig a decent burrow – having both the inclination and the anatomical wherewithal to do so – would constitute a survival advantage for that animal and its offspring. Birds build wonderful nests, affording similar privileges and similar advantages in survival. Some chimpanzees use stones to crack nuts and sticks to fish termites out of their nests, allowing them to access an otherwise inaccessible food resource. All of these activities – den-digging, nest-weaving, using tools to get at food – provide survival advantages to the animals engaging in them. We wouldn't baulk at the idea that their anatomy and physiology might have evolved to support those activities, that features which might help them to be better diggers, weavers or termite-fishers would be selected for, while features which impeded such activities might be weeded out by natural selection.

When we turn the same lens on our species, then, we shouldn't be surprised to find that our own structure and function has been influenced by the things that our ancestors did and made, which provided them with a survival advantage. In fact, it's even more prominent in the evolution of humans than in that of any other animal, because we have such a profound effect on the environments we exist in. The environment shapes us, but we're also shaping the environment that shapes us.

Our hands have been shaped by the technology we create, because that technology has played a part in our survival and our success as a species. In the same way, our brains have been sculpted by ways of thinking and doing which proved advantageous to our ancestors.

Our own technology and culture becomes part of the environment we live in. This is the era of the world wide web, where we can share ideas much more quickly and widely than ever before – just imagine how our own technology will be shaping human brains in the future. The brain of a newborn baby still has so much developing to do, much of it involving pruning back connections and consolidating certain pathways, and this happens in a social and cultural context. It's impossible for our minds *not* to be shaped by culture and technology – that's how they're designed to develop. Remember Michael Tomasello's aphorism: a human baby is born expecting culture just as a fish is born expecting water. We might, quite reasonably, worry that too much 'screen-time' detracts from other activities – from real rather than virtual social interactions, from physical activities, from spending time outdoors. We may worry in particular about the effects of all that screen-time on our kids. I think it's absolutely right to be cautious, but we shouldn't be afraid of the technology itself – it's part of our environment, and our children are growing up in a different environment than their parents did. It's sobering to remember that people had similar worries about books when printing and publishing took off. Thank goodness the adaptability of the human brain means that our children will pick up new knowledge, skills and understanding – new ways of thinking – which will enable them to adapt to this changing cultural environment. We shouldn't think that adult brains are 'set solid' either: neural plasticity, including the ability to create new connections, is a lifelong feature of our brains.

HUMAN UNIQUENESS

Thinking about brains focuses our attention on the part of us which we really do feel to be unique. There's no doubt that we love to think of ourselves as special, but what does this really mean? In a broad sense, *Homo sapiens* is unique as a species; we have a unique combination of traits like habitually walking around on two legs, very dextrous hands and massive brains. But isn't every species unique? That's the whole point, surely?

Numerous hypotheses, which have assumed particular features to be uniquely human, have been proved wrong, or at least not quite right, on closer inspection. The obstetric dilemma hypothesis suggested that the female human pelvis was victim to a tug-of-war between the demands of bipedalism and birthing a big-brained infant. The expensive tissue hypothesis suggested that humans had uniquely small guts for an ape, and this does not appear to be entirely true. Even regularly walking around on the ground doesn't seem to be unique to just our own, hominin sprig of the ape family tree – *Oreopithecus* was doing it too, long before any hominins came along.

Many of our features which have been considered to be unique turn out to be differences in degree: quantitative rather than qualitative differences. Other living apes also walk on two legs occasionally, it's just that we do it habitually. Primates have large brains compared with other mammals, but we've taken it a bit further. These differences of degree make sense, of course, if you consider that the human body *has* evolved.

It's not a new creation, it didn't just appear out of nothing, and neither did any of the anatomical components which make it up. As our minds seem to be qualitatively different from those of our closest living relatives, this prompts the question: when did this change happen? It's likely that the 'human mind', as we know it, emerged gradually. In the same way in which the whole 'human package', including bipedalism, long, flexible waists, low shoulders, big brains, etc, did not arrive overnight but appeared in a mosaic or piecemeal fashion, the minds of our ancestors would not have suddenly arrived at full human consciousness.

In modern biology, the idea of a linear progression, starting with 'lower' animals and progressing to 'higher' ones, has been replaced by the concept of a luxuriantly branching tree of life. But the idea of a linear *scala naturae*, with humanity as its final destination, is very tenacious. Particularly because we do have such an impact on the world around us, it's very tempting indeed to believe that we are, somehow, the pinnacle of evolution. In the early twentieth century, evolutionary biologists were still writing in this vein, portraying human evolution as a natural extension of trends in primate evolution, as an almost inevitable and progressive development. But by the 1950s, fossil discoveries were helping to reveal a different history, one in which unpredictable changes in the environment affected the way species evolved. Human evolution was seen to involve contingency and serendipity.

In the second half of the twentieth century, as more hominin fossils were discovered and our recent family tree became populated with more and more species, it became very clear that our distinctive features had not arrived as a 'package'. Walking on two legs came along millions of years before brain enlargement. But scholars of human evolution still sought a defining *moment* for the true origins of humanity, an 'adaptive shift' which would set our ancestors off in a brand new direction. In the 1960s, the shift was suggested to be hunting and eating meat, and 'Man the Hunter' strode bravely onto the scene: predation was the change, the new behaviour, which defined humanity. Nowadays, such ideas seem quaint – a search for a mythical holy grail. If the path to hell is paved with good intentions, the path to humanity is paved with numerous, some-times very subtle, shifts in behaviour, physiology, anatomy – and consciousness. It is foolish to look for *one* thing which takes a group of animals off in a new evolutionary direction. Apart from anything else, that 'direction' is only apparent with hindsight. It may be possible to iden-tify key innovations, but it's still very unlikely that *one* change explains most of the differences between us and our closest living relatives.

Like any other species, *Homo sapiens* represents a variation on a theme, as an animal, a vertebrate, a mammal and an ape. But many species have something which sets them apart, a unique and unusual feature which has developed to such an extent that it makes those animals stand out from the crowd. In peacocks, it's the extraordinary, long and resplendent tail of the male. In humans, surely it's our incred-ibly large brains, and what we do with them. Many of our bodily func-tions and the anatomy that supports them – from the digestive processes going on in our guts, to the way our hearts pump blood around the body, and the way in which our eyes sense different wavelengths of light – are incredibly similar to what we see in other apes, even other mammals. But our brains are *so* large, and so full of knowledge and thoughts and feelings that we're fairly sure we must be far beyond the reach of any other animal. Surely our minds are unique?

As far as we know, our type of cognition *is* unique among living animals, but even while we may be sure that whole package is something which seems to be uniquely human, if we start to analyse various aspects of our psychology we find that the differences between our minds and those of other animals are in fact differences of degree, rather than complete departures, or absolute differences, from the mental states of other creatures. Even chimpanzees seem to have some 'theory of mind'; some grasp that other individuals possess minds which contain thoughts, desires and aspirations.

There are perhaps differences in how deeply that theory of mind works in chimpanzees compared with humans. But this means that, once again, we're left looking at a difference of degree instead of an absolute difference. We're not as different, 'as unique', as we might have thought. This is still quite an unsettling thought, as it was, after all, one of the major reasons that many people found (and in some unenlightened places, still find) Darwin's description and explanation of evolution so unbelievable. We're not a unique creation set apart from the rest of nature, no matter how much we'd love that to be true.

In 1860, Thomas Henry Huxley, 'Darwin's Bulldog', found himself arguing with his opponent, Archbishop Samuel Wilberforce, in an impromptu debate about evolution at the Oxford University Museum of Natural History. During the exchange, Wilberforce laid into evolution and asked Huxley on which side of his family he claimed descent from a monkey – on his grandmother's side or his grandfather's? Accounts vary as to exactly what happened next, but according to one version, Huxley muttered 'The Lord hath delivered him into my hands', and then stood up to offer this decisive blow: 'I would rather be the offspring of two apes than be a man and afraid to face the truth.'

Are we still uncomfortable about thinking of ourselves as apes, as mammals, as vertebrates? Because that's what we are. A hundred and thirty four years after Huxley jousted with Wilberforce in Oxford, the biological anthropologist Matt Cartmill wrote, 'to seek to show that all things human are prefigured or paralleled in the lives and adaptations of our fellow animals . . . is at bottom to doubt the reality of the moral boundary that separates people from the beasts. Whether we fear or welcome the dissolution of that boundary is the real issue.'

FINDING OUR PLACE

We are apes that have done very well, I'm not denying that. It's extraordinary that a species of ape has managed to be quite so successful. We create wonderful things, including art, music and literature, as well as technology which improves our chances of survival, reproduction and longevity.

But just as we, as individuals, are not going to live forever (however hard that is to stomach), our species is not going to be here for all eternity, either. We might have slowed evolution down a bit, taken the edge off the scythe of the grim reaper we know as natural selection, in developed countries, at least, but even in such privileged places where each baby has a very good chance of surviving to adulthood, there will be differences in how many children couples have, and those differences will change the frequency of genes in the population. Slowly, perhaps, but it's still evolution.

It's possible that no significant changes will occur to us, at least anatomically, while we're in this exalted state where we can control and maintain the stability of the environment we live in. We could become living fossils, like horseshoe crabs and coelacanths, while other species rise and fall around us. We have given ourselves a real fighting chance of surviving local cataclysms by spreading right across the globe, in vast numbers. But ultimately, it's likely that a catastrophic change to our environment, which, let's face it, could even be of our own making, will drastically change the rules of the game. (Just remember the dinosaurs and that Chicxulub asteroid.) At that point, our species might be extinguished, or it could be reduced to a few small populations hanging on in places which are still just habitable. In those refugia, natural selection would sharpen up its scythe and get to work, and the effects of genetic drift could also be profound. In such circumstances, the future of humanity could look very different from its present incarnation.

I'm not going to polish up my crystal ball to try to predict the future for our species – there is too much unpredictability in evolution and in the galaxy for that – but I will make some predictions about how humans won't change in the near future: we won't grow extra, fully functioning fingers or toes; the pentadactyl pattern is too deeply embedded in our genomes now to make that at all likely. We won't grow wings, or extra legs, for the same reason. As long as we retain some technology to keep us warm in cold places (shelters, clothes, fire), we won't grow furry again – unless that becomes, inexplicably, something which is considered to be very attractive.

It's as hard to predict where our evolutionary destiny lies as it would have been to predict, 66 million years ago, that some of the mammals who hid from the dinosaurs would have evolved into monkeys, that some of those would have evolved into apes, and that some apes would have become habitual terrestrial bipeds – very good with their hands and very clever. I don't think the happenchance and contingency (which is still there, albeit channelled by constraints) of evolution should make us feel inconsequential or insignificant. For me, well, I feel extraordinarily lucky to be here. Just imagine, for a moment, how easy it would have been *not* to be here.

There are innumerable points in the evolutionary history of our species where circumstances could have taken a different turn, or where the lineage we can trace back through the tree of life might have been pruned short rather than continuing on to what is currently the periphery of the bush. As an individual, the chances of *your* conception are vanishingly small. Your beginning relied on that *one* sperm among millions making it to the egg which one of your mother's ovaries had released that month. It could so easily have been different.

While each of us can feel lucky to be alive, we also have a heavy burden to bear, the knowledge of a weighty responsibility. We have deep theory of mind which allows us to gauge others' intentions and aims. We can grasp and communicate abstract ideas. We also have a sense of the impact of our own actions, not only individually and directly, but collectively and globally, informed by the investigative urge we call science.

Through studying evolution and embryology we find our place in the world. We are part of the story of life, connected with other species, past and present, in a very real way. After that chance moment of conception, your own epigenesis began: form appeared where there was no form before, governed by a code written in an ancient language, generating fleeting echoes of past lives and distant cousins as the drama unfolded. You are part, not of a great chain of being – a *scala naturae* – but of the huge, copiously branching *arbor naturae*, the tree of life.

Because we are such unusual animals, because we are conscious of our own impact on others, we bear an obligation not only to each other, but to the other twigs on that tree of life as well. Accepting our place within the natural world, not outside it, surely means that we must shoulder this responsibility and work towards a sustainable future, for ourselves and for all the other species on this planet.

ACKNOWLEDGEMENTS

&

FURTHER READING

ACKNOWLEDGEMENTS

I have wanted to write a book about embryology and evolution for years, and Susan and Jon Watt of Heron Books were kind enough to let me regale them with stories about development, common ancestors and family trees – and then brave enough to publish the book! I'm enormously grateful to both of them for their encouragement and their enthusiasm for what could be seen as a rather niche subject.

Many of the ideas in the book have flowed from meetings with numerous colleagues, often while filming television programmes. I'm very grateful to the producers and production teams of the various programmes and series which have allowed me to indulge my passion for evolutionary biology and physical anthropology, to meet inspirational researchers in the course of making these programmes, and to share this exciting field of science with a wide audience. In particular, I would like to thank the production team of *Origins of Us* (BBC2, 2011), including: Mags Lightbody and Davina Bristow (Assistant Producers); Dave Stewart, Matt Dyas and Paul Olding (Producer/Directors); Zoe Heron (Series Producer); Sacha Baveystock (Executive Producer); the Executive Producer of *Prehistoric Autopsy* (BBC2, 2012), Jane Aldous; and the production team of *Horizon: What makes us human?*, including Chris Pitt (Researcher), Toby Macdonald (Producer/Director) and Series Editor, Aidan Laverty. And huge thanks to Kim Shillinglaw, Commissioning Editor for Science and Natural History for the BBC.

I wish to thank all those who have generously shared their research and their ideas with me; some colleagues are mentioned in the text, including: Jeff Lichtman of Harvard University; Geoff Bird of Birkbeck

College; Denis Duboule of Geneva University; Anna Barney of Southampton University and Sandra Martelli of University College London; Roy Cockel of Selly Oak Hospital; Holly Dunsworth of Rhode Island University; Herman Pontzer of City University of New York; Anna Warrener of Harvard University. And then there are many other colleagues who have shared their biological, anatomical, evolutionary and epigenetic expertise with me, in conversations and through correspondence, and to whom I am also extremely grateful, including: Leslie Aiello of the Wenner-Gren Foundation for Anthropological Research; Sir Patrick Bateson and David Chivers of the University of Cambridge; Jeremy Pritchard, Susannah Thorpe and Bryan Turner of the University of Birmingham. As a medical student, I was lucky to learn from inspirational teachers of anatomy and embryology, including Edward Evans, Bernard Moxham and Richard Newell. But I am particularly indebted to Bob Presley, the anatomist who supervised my undergraduate anatomy dissertation project and who introduced me to the world of comparative anatomy and embryology.

Several colleagues kindly read and commented on early drafts of the book, and I am enormously grateful to them for their invaluable insights, suggestions and gentle corrections. These generous-hearted people include: Robin Crompton, Holly Dunsworth, Colin Groves, Desmond Morris, Mark Pallen, Chris Stringer and Paolo Viscardi. Thanks also to Jon Watt, my editor, for his perspicacity and for helping me to find and develop the narrative thread in the book. Thank you, too, to my copy editor, Helena Caldon, who smoothed out many wrinkles.

Last, but not least, I couldn't have written this book while looking after a new baby and a three-year-old, without the incredible support of my husband. Thank you, Dave.

FURTHER READING

General

Here are a few excellent books and online resources dealing with general themes of embryology and evolution, as well as the history of science in this area.

Books:

Cobb, M. (2007). *The egg and sperm race: the seventeenth-century scientists who unravelled the secrets of sex, life and growth.* London; New York: Pocket Books.

Correia, C. P. (1998). *The Ovary of Eve: egg and sperm and preformation.* Chicago: University of Chicago Press.

Gould, S. J. (1977). *Ontogeny and Phylogeny.* Cambridge, Massachusetts: Belknap Press of Harvard University Press.

Kent, G. C. & Carr, R. K. (2001). *Comparative anatomy of the vertebrates.* Boston: McGraw Hill.

Lieberman, D. (2011). *The evolution of the human head.* Cambridge, Massachusetts: Belknap Press of Harvard University Press.

Sadler, T. W. (2012). *Langman's Medical Embryology.* Philadelphia: Lippincott, Williams and Wilkins.

Shubin, N. (2008). *Your inner fish: a journey into the 3.5-billion-year history of the human body.* New York, Pantheon Books.

Wolpert, L. (2008) *The Triumph of the Embryo.* Mineolla, NY: Dover Publications.

Online and mobile apps:

An online embryology course developed by the Universities of Fribourg, Lausanne & Berne, Switzerland, available in French, German, English and Dutch: www.embryology.ch

The Embryo Project Encyclopaedia, from Arizona State University, an online repository of 'found objects', descriptive articles and essays about embryology and its history: www.emo.asu.edu

A beautiful guide to embryological development and changes during pregnancy: lifeinthewombapp.com

Beginnings

A brief history of ideas

Cobb, M. (2012). An amazing 10 years: the discovery of egg and sperm in the 17th century. *Reproduction in Domestic Animals*, 47: 2–6.

Gould, S. J. (1977). *Ontogeny and Phylogeny.* Cambridge, Massachusetts: Belknap Press of Harvard University Press.

The beginning of you

Do check out this stunning visual guide to pregnancy: lifeinthewombapp.com

Heads and brains

The first head

On the simplification of sea squirts:

Holland, N. D. & Chen, J. (2001). Origin and early evolution of the vertebrates: new insights from advances in molecular biology, anatomy and palaeontology. *BioEssays*, 23: 142–151.

On the discovery of *Haikouella*:

Chen, J.–.Y., Huang, D.–Y., Li, C.–W. (1999). An early Cambrian craniate-like chordate. *Nature*, 402: 518–522.

Richard Dawkins discusses our relationship to starfish and our place in the great tree of life in his book, *The Ancestor's Tale* (2004).

For more on acorn worms and the origin of chordates:

Brown, F. D., Prendergast, A., Swalla, B. J. (2008). Man is but a worm: chordate origins. *Genesis*, 46: 605–613.

Gerhart, J., Lowe, C., Kirschner, M. (2005). Hemichordates and the origin of chordates. *Current Opinion in Genetics & Development*, 15: 461–467.

Lacalli, T. C. (2010). The emergence of the chordate body plan: some puzzles and problems. *Acta Zoologica* (Stockholm), 91: 4–10.

Link to BBC *Life* starfish film on YouTube:

www.youtube.com/watch?v=kGMCaTwkKrc

Ancient and embryonic brains

For more on neurulation and sonic hedgehog, try the Embryo Project Encyclopaedia: www.emo.asu.edu

For more on brains in different animals:

Kent, G. C. & Carr, R. K. (2001). *Comparative anatomy of the vertebrates.* Boston: McGraw Hill.

Mapping the human brain

For more on brain function and the anatomy of the head and neck generally:

Lieberman, D. (2011). *The evolution of the human head.* Cambridge, Massachussetts: Belknap Press of Harvard University Press.

For more on the extraordinary story of Phineas Gage and his injury:

Ratiu, P. *et al.* (2004). The tale of Phineas Gage, digitally remastered. *Journal of Neurotrauma*, 21: 637–643.

On the connections in the human brain:

Drachman, D. A. (2005). Do we have brain to spare? *Neurology*, 64: 2004–2005.

For more on the Brainbow:

Livet, J. *et al.* (2007). Transgenic strategies for combinatorial expression of fluorescent proteins in the nervous system. *Nature*, 450: 56–63.

For more on brain imaging:

Raichle, M. E. (2008). A brief history of human brain mapping. *Trends in Neurosciences*, 32: 118–126.

Mirror neurons

Catmur, C. *et al.* (2008). Through the looking glass: counter-mirror activation following incompatible sensorimotor learning. *European Journal of Neuroscience*, 28: 1208–1215.

Heyes, C. (2010). Where do mirror neurons come from? *Neuroscience and Biobehavioural Reviews*, 34: 575–583.

Ramachandran, V. S. (2000). Mirror neurons and imitation learning as the driving force behind 'the great leap forward' in human evolution. www.edge.org

Ramachandran, V. S. (2011). *The Tell-Tale Brain: Unlocking the Mystery of Human Nature.* London: Windmill Books. Chapter 4: The neurons that shaped civilization.

The huge human brain

For a great discussion of EQ and human evolution, see:

Lieberman, D. (2011). *The evolution of the human head.* Cambridge, Massachussetts: Belknap Press of Harvard University Press.

On the visual cortex of chimpanzees:

Holloway, R. L., Broadfield, D. C., Yuan, M. S. (2003). Morphology and histology of chimpanzee primary visual striate cortex indicate that brain reorganization predated brain expansion in early hominid evolution. *The Anatomical Record*, 273A: 594–602.

Wood, J. N. & Grafman, J. (2003). Human prefrontal cortex: processing and representational perspectives. *Nature Review Neuroscience*, 4: 139–147.

On the history of research into chimpanzee psychology:

Call, J. & Tomasello, M. (2008). Does the chimpanzee have a theory of mind? 30 years later. *Trends in Cognitive Science*, 12: 187–192.

On the effect of the expanding human brain on the shape of the skull:

Lieberman, D. E., Pearson, O. M., Mowbray, K. M. (2003). Basicranial influence on overall cranial shape. *Journal of Human Evolution*, 38: 291–315.

For more on the evolution of modern human skull shape:

Lieberman, D. E., McBrateney, B. M., Krovitz, G. (2002). The evolution and development of cranial form in *Homo sapiens*. *Proceedings of the National Academy of Sciences*, 99: 1134–1139.

Skulls and senses

Neural crest and the origin of the skull

On searching for the evolutionary origins of new structures, and the comment that 'Morphological structures . . . do not simply arise from the dust of the earth':

Braun, C. B. & Northcutt, R. G. (1997). The lateral line system of hagfishes. *Acta Zoologica* (Stockholm), 78: 247–268.

On gene duplication and new jobs for genes in evolution:

Holland, L. Z., Laudet, V., Schubert, M. (2004). The chordate amphioxus: an emerging model organism for developmental biology. *Cellular and Molecular Life Sciences*, 61: 2290–2308.

On the not-quite-neural-crest cells of lancelets:

Shimeld, S. M., Holland, N. D. (2005). Amphioxus molecular biology: insights into vertebrate evolution and developmental mechanisms. *Canadian Journal of Zoology*, 83: 90–100.

On Julia Platt and her discovery that neural crest cells contributed to the developing skull:

Landacre, F. L. (1921). The fate of the neural crest in the head of the urodeles. *The Journal of Comparative Neurology*, 33: 1–43.

Zottoli, S. J. & Seyfarth, E.-A. (1994). Julia B. Platt (1857–1935): Pioneer comparative embryologist and neuroscientist. *Brain, behaviour and evolution*, 43: 92–106.

Making a skull

Charles Darwin wrote about his 'law of embryonic resemblance' in *On the Origin of Species*, published in 1859. It's worth a read. There's also a wonderful online resource: darwin-online.org.uk

Skull shapes

On skull deformation:

Ayer, A. *et al.* (2010). The sociopolitical history and physiological underpinnings of skull deformation. *Neurosurgical Focus*, 29: 1–6.

Tubbs, R. S., Salter, E. G., Oakes, W. J. (2006). Artificial Deformation of the Human Skull. *Clinical Anatomy*, 19: 372–377.

On the deformed juvenile skull known as the 'starchild':

Novella, S. (2006). The Starchild Project, *The New England Sceptical Society*: www.theness.com/index.php/the-starchild-project

Smelling

The study showing lack of awareness of loss of smell following brain injury:

Callahan, C. D. & Hinkebein, J. H. (2002). Assessment of anosmia after traumatic brain injury. *Journal of Head Trauma Rehabilitation*, 17: 251–256.

On smell genes in sea lampreys:

Libants, S. *et al.* (2002). The sea lamprey *Petromyzon marinus* genome reveals the early origin of several chemosensory receptor families in the vertebrate lineage. *BMC Evolutionary Biology*, 9:180.

On the hypothesis linking loss of smell genes to the development of colour vision in primates:

Gilad, Y. *et al.* (2004). Loss of olfactory receptor genes coincides with the acquisition of full trichromatic vision in primates. *PLoS Biology*, 2: 120–125.

Matsui, A., Go, Y., Niimura, Y. (2010). Degeneration of olfactory receptor gene repertoires in primates: no direct link to full trichromatic vision. *Molecular Biology & Evolution*, 27: 1192–1200.

Seeing

For more on eye evolution:

Lamb, T. D., Collin, S. P., Pugh, E. N. (2007). Evolution of the vertebrate eye: opsins, photoreceptors, retina and eye cup. *Nature Reviews: Neuroscience*, 8: 960–975.

On the study looking at the genes expressed in the developing lancelet's eye:

Vopalensky, P. *et al.* (2012). Molecular analysis of the amphioxus frontal eye unravels the evolutionary origin of the retina and pigment cells of the vertebrate eye. *Proceedings of the National Academy of Sciences*, 109: 15383–15388.

On colour vision and light receptors in vertebrates:

Bowmaker, J. K. (1998). Evolution of colour vision in vertebrates. *Eye*, 12: 541–547.

On colour vision in primates and red bottoms:

Surridge, A. K., Osorio, D., Mundy, N. I. (2003). Evolution and selection of trichromatic vision in primates. *Trends in Ecology and Evolution*, 18: 198–205.

On the diets of early primates (and a refreshing take on human evolution):

Cartmill, M. (2012). Primate origins, human origins, and the end of higher taxa. *Evolutionary Anthropology*, 21: 208–220.

On gorillas attending to the direction of eye gaze:

Mayhew, J. A. (2013). Attention cues in apes and their role in social play behaviour of Western Lowland Gorillas (*Gorilla gorilla gorilla*). PhD thesis, University of St Andrews.

On human infants attending to eye gaze, and the cup experiment to test if children will detect deception:

Freire, A., Eskritt, M., Lee, K. (2004). Are eyes windows to a deceiver's soul? Children's use of another's eye gaze cues in a deceptive situation. *Developmental Psychology*, 6: 1093–1104.

On comparing attention to eye gaze in human infants and other great apes:

Tomasello, M. *et al.* (2007). Reliance of head versus eyes in the gaze following of great apes and human infants: the cooperative eye hypothesis. *Journal of Human Evolution*, 52: 314–320.

Speech and gills

A U-shaped bone and butterfly-shaped cartilage

On the discovery of the Kebara Neanderthal:

Bar-Yosef, O. *et al.* (1992). The excavations in Kebara Cave, Mt Carmel. *Current Anthropology*, 33: 497–550.

On the hypothesis that the human tongue is supplied more richly with nerves than that of other apes, and that hypoglossal canal size might be a useful clue:

DeGusta, D., Gilbert, W. H., Turner, S. P. (1999). Hypoglossal canal size and hominid speech. *Proceedings of the National Academy of Sciences*, 96: 1800–1804.

Kay, R. F., Cartmill, M., Balow, M. (1998). The hypoglossal canal and the origin of human vocal behaviour. *Proceedings of the National Academy of Sciences*, 95: 5417–5419.

On the hyoid bones from the Sima de los Huesos:

Martinez, I. *et al.* (2007). Human hyoid bones from the middle Pleistocene site of the Sima de los Huesos (Sierra de Atapuerca, Spain). *Journal of Human Evolution*, 54: 118–124.

On how articulate a Neanderthal might have been:

Boe, L.-J. *et al.* (2002). The potential Neanderthal vowel space was as large as that of modern humans. *Journal of Phoentics*, 30: 465–484.

Laryngeal myths

On the lack of increased risk of choking with a descended larynx:

Aiello, L. (2002). Fossil and other evidence for the origin of language. *Evolution of Language 4th International Conference*, Harvard University.

On the larynx being squeezed down in the neck:

Lieberman, D. (2011). The evolution of the human head. Cambridge, Massachusetts: Belknap Press of Harvard University Press.

Lieberman, D. E. *et al.* (2001). Ontogeny of postnatal hyoid and larynx descent in humans. *Archives of Oral Biology*, 46: 117–128.

On the different vowel sounds of men and women:

de Boer, B. (2010). Investigating the acoustic effect of the descended larynx with articulatory models. *Journal of Phonetics*, 38: 679–686.

On other animals with low larynges:

Fitch, W. T. & Reby, D. (2001). The descended larynx is not uniquely human. *Proceedings of the Royal Society of London, B* 268: 1669–1675.

Deep-voiced males

On roaring stags:

Clutton-Brock, T. H. & Albon, S. D. (1979). The roaring of red deer and the evolution of honest advertisement. *Behaviour*, 69: 145–170.

On the effect of low-voiced men on women:

Smith, D. S. *et al.* (2012). A modulatory effect of male voice pitch on long-term memory in women: evidence of adaptation for mate choice? *Memory & Cognition*, 40: 135–144.

The dating game at the University of Pittsburgh:

Puts, D. A., Gaulin, S. J. C., Verdolini, K. (2006). Dominance and the evolution of sexual dimorphism in human voice pitch. *Evolution and Human Behavior*, 27: 283–296.

Jaw joints and ossicles

On Morganucudon's second jaw joint:

Kermack, K. A., Mussett, F., Rigney, H. W. (1973). The lower jaw of *Morganucodon*. *Zoological Journal of the Linnean Society*, 53: 87–175.

The origins of muscles in the ear and on the face as gill muscles:

Diogo, R. (2008). Comparative anatomy, homologies and evolution of the mandibular, hyoid and hypobranchial muscles of bony fish and tetrapods: a new insight. *Animal Biology*, 58: 123–172.

Takechi, M. & Kuratani, S. (2010). History of studies on mammalian middle ear evolution: a comparative morphological and developmental biology perspective. *Journal of Experimental Zoology B*, 314: 1–17.

Von Baer and genetics

The quote about pharyngeal or branchial arches is from one of the most popular undergraduate embryology textbooks:

Sadler, T. W. (2012). *Langman's Medical Embryology*. Philadelphia: Lippincott, Williams and Wilkins.

Spines and segments

Fruit flies and the origin of spines

For more on Thomas Hunt Morgan and his Nobel-prize-winning work, search: NobelPrize.org

On *Hox* genes:

Myers, P. Z. (2008). *Hox* genes in development: the *Hox* code. Scitable, Nature Education.

Pearson, J. C., Lemons, D., McGinnis (2005). Modulating Hox gene functions during animal body patterning. *Nature Review Genetics*, 6: 893–904.

On our ancient common ancestor with fruit flies:

Danchin, E. G., Pontarotti, P. (2004). Statistical evidence for a more than 800-million-year-old evolutionarily conserved genomic region in our genome. *Journal of Molecular Biology and Evolution*, 59: 587–597.

On human *Hox* genes:

Holland, P. W. H., Booth, H. A. F., Bruford, E. A. (2007). Classification and nomenclature of all human homeobox genes. *BMC Biology*, 5: 47.

More about Denis Duboule:

Richardson, M. K. (2009). The Hox complex: an interview with Denis Duboule. *International Journal of Developmental Biology*, 53: 717–723.

Embryological development of vertebrae

For more information about spina bifida, try NHS National Genetics and Genomics Education Centre > Genetics Conditions > Neural tube defects: www.geneticseducation.nhs.uk?genetic-conditions-54/688-neural-tube-defects-new

Hox genes in the development of the spine:

Carapuco, M. *et al.* (2005). *Hox* genes specify vertebral types in the presomitic mesoderm. *Genes and Development*, 19: 2116–2121.

Slipped discs and faulty facets

On the loss of notochordal cells from intervertebral discs:

Hunter, C. J., Matyas, J. R., Duncan, N. A. (2003). The notochordal cell in the nucleus pulposus: a review in the context of tissue engineering. *Tissue Engineering*, 9: 667–677.

On lower back pain:

Cohen, S. P. & Raja, S. N. (2007). Pathogenesis, diagnosis, and treatment of lumbar zygapophysial (facet) joint pain. *Anesthesiology*, 106: 591–614.

Manchikanti, L. *et al.* (2004). Prevalence of facet joint pain in chronic spinal pain of cervical, thoracic and lumbar regions. *BMC Musculoskeletal Disorders*, 5: 15.

Long lumbar spines

On the evolution of monkeys and ancient long, flexible spines:

Harrison, T. (2013). Catarrhine Origins. In Begun, D. R. (ed.) *A Companion to Paleoanthropology*, pp376–396.

On the mysterious loss of a tail in apes:

Larson, S. G. & Stern, J. T. (2006). Maintenance of above-branch balance during primate arboreal quadrupedalism: coordinated use of forearm rotators and tail motion. *American Journal of Physical Anthropology*, 129: 71–81.

On the lumbar spine in fossil hominins:

Lovejoy, C. O. (2005). The natural history of human gait and posture, Part 1: spine and pelvis. *Gait and posture*, 21: 95–112.

A straight-backed cousin

On measuring the degree of lordosis in ancient spines:

Been, E., Gomez-Olivencia, A., Kramer, P. A. (2012). Lumbar lordosis of extinct hominins. *Americam Journal of Physical Anthropology*, 147: 64–77.

Ribs, lungs and hearts

Thoracic shapes

On the shape of apes' chests:

Kagaya, M., Ogihara, N., Nakatsukasa, M. (2008). Morphological study of the anthropoid thoracic cage: scaling of thoracic width and an analysis of rib curvature. *Primates*, 49: 89–99.

On anatomy and locomotion in apes:

Larson, S. G. (1998). Parallel evolution in the hominoid trunk and forelimb. *Evolutionary Anthropology*, 6: 87–99.

Fossil ribs

On evolutionary changes in hominin chest shape:

Haile-Selassie, Y. *et al.* (2010). An early *Australopithecus afarensis* postcranium from Woranso-Mille, Ethiopia. *Proceedings of the National Academy of Science*, 107: 12121–12126.

Schmid, P. *et al.* (2013). Mosaic morphology in the thorax of *Australopithecus sediba*. *Science*, 340: 1234598.

Wong, K. (2013). Is *Australopithecus sediba* the most important human ancestor discovery ever? *Scientific American*, April 23.

On the likelihood of hominins evolving from an ancestor with a small ape with a chest shape similar to that of gibbons or spider monkeys:

Lovejoy, C. O. *et al.* (2009). The pelvis and femur of *Ardipithecus ramidus*: the emergence of upright walking. *Science*, 326: p71e5.

Stern, J. T. *et al.* (1980). An electromyographic study of the pectoralis major in Atelines and Hylobates with special references to the evolution of a pars clavicularis. *American Journal of Physical Anthropology*, 20: 498–507.

Neanderthal chests

Gomez-Olivencia *et al.* (2009). Kebara 2: new insights regarding the most complete Neanderthal thorax. *Journal of Human Evolution*, 57: 75–90.

Making lungs

Suzuki *et al.* (2010). The mitochondrial phylogeny of an ancient lineage of ray-finned fishes (Polypteridae) with implications for the evolution of body elongation, pelvic fin loss and craniofacial morphology in Osteichthyes. *BMC Evolutionary Biology*, 10: 21.

Guts and yolk sacs

A fantastic voyage

Using pillcams to investigate patients' intestines at Selly Oak Hospital:
Hudson, J. (2002). Hope at end of fantastic voyage. *Birmingham Post*, May 24.

On the 'un-uniqueness' of human guts

On the lack of a link between brain and gut size in different animals:
Hladik, C. M. & Pasquet, P. (2003). Reply to: Kaufman, J. A. (2003). On the expensive tissue hypothesis: independent support from highly encephalised fish. *Current Anthropology*, 44: 705–707.

The study on brain size and fat in humans and other animals:
Navarrete, A., van Schaik, C. P., Isler, K. (2011). Energetics and the evolution of human brain size. *Nature*, 480: 91–93.

On the switch to a high-energy diet providing the means for growing bigger brains and bodies:
Pontzer, H. (2012). Ecological energetics in early *Homo*. *Current Anthropology*, 53, S6: S346–S358.

On lactase persistence and the ability to digest milk into adulthood:
Ingram, C. J. E. *et al.* (2009). Lactose digestion and the evolutionary genetics of lactase persistence. *Human Genetics*, 124: 579–591.

Gonads, genitals and gestation

Lumps, bumps and tubes

For more on the anatomy and function of the external genitalia:
Berman, J. R. & Bassuk, J. (2002). Physiology and pathophysiology of female sexual function and dysfunction. *World Journal of Urology*, 20: 111–118.

O'Connell, H. E. (2005). Anatomy of the clitoris. *Journal of Urology*, 174: 1189–1195.

On Mullerian and Wolffian ducts:
Brian, B. H. & Bloom, D. A. (2002). Wolff and Muller: fundamental eponyms of embryology, nephrology and urology. *The Journal of Urology*, 168: 425–428.

Mind the bollocks: the incredible migrating testis (and ovary)

Ivell, R. (2007). Lifestyle impact and the biology of the human scrotum. *Reproductive Biology & Endocrinology*, 5: 15.

On the weight of human, chimpanzee and gorilla testes:

Dixson, A. F. & Anderson, M. J. (2004). Sexual behaviour, reproductive physiology and sperm competition in male mammals. *Physiology & Behaviour*, 83: 361–371.

Penises, clitorises and orgasm

On penis-like clitorises:

Place, N. J. & Glickman, S. E. (2004). Masculinization of female mammals: lessons from nature. *Advances in Experimental Medical Biology*, 545: 243–253.

For more on penile and clitoral erection:

Dean, R. C. & Lue, T. F. (2005). Physiology of penile erection and pathophysiology of erectile dysfunction. *Urologic Clinics of North America*, 32: 379–395.

Gragasin, F. S. *et al.* (2004). The neurovascular mechanism of clitoral erection: nitric oxide and cGMP-stimulated activation of BKCa channels. *Federation of American Societies for Experimental Biology*, 18: 1382–1391.

On oxytocin:

Borrow, A. P. & Cameron, N. M. (2012). The role of oxytocin in mating and pregnancy. *Hormones and Behavior*, 61: 266–276.

On the mysteries of the female orgasm:

Colson, M.-H. (2010). Female orgasm: myths, facts and controversies. *Sexologies*, 19: 8–14.

Getting it on

On lap dancers earning more at ovulation:

Miller, G., Tybur, J. M., Jordan, B. D. (2007). Ovulatory cycle effects on tip earnings by lap dancers: economic evidence for human estrus? *Evolution and Human Behavior*, 28: 375–381.

On sex among pygmy chimpanzees or bonobos:

de Waal, F. B. M. (1995). Bonobo sex and society. *Scientific American*, March.

A tight squeeze

The importance of social learning to us humans:

Sterelny, K. (2011). From hominins to humans: how *sapiens* became behaviourally modern. *Philosophical Transactions of the Royal Society B*, 366: 809–822.

Introducing the 'obstetric dilemma':

Rosenberg, K. R. (1992). The evolution of modern human childbirth. *Yearbook of Physical Anthropology*, 35, 89–124.

Rosenberg, K. R. & Trevathan, W. (2002). Birth, obstetrics and human evolution. *British Journal of Obstetrics & Gynaecology*, 109: 1199–1206.

On the new challenges to the obstetric dilemma:

Dunsworth, H. M. *et al.* (2012). Metabolic hypothesis for human altriciality. *Proceedings of the National Academy of Science*, 109: 15212–15216.

Meeting Holly Dunsworth to talk about the obstetric dilemma (from 28.00 minutes) BBC2 *Horizon: What makes us human?* www.youtube.com/watch?v=AqK6eE51Ctk

Wells, J. C. K., DeSilva, J. M., Stock, J. T. (2012). The obstetric dilemma: an ancient game of Russian roulette, or a variable dilemma sensitive to ecology? *Yearbook of Physical Anthropology*, 55: 40–71.

On the difficulty of tracking down the trigger for human birth:

Plunkett, J. *et al.* (2011). An evolutionary genomic approach to identify genes involved in human birth timing. *PLoS Genetics*, 7: e1001365.

A general introduction to human reproduction, and the context of birth:

Ellison, P.T. (2001). *On fertile ground*. Cambridge, MA: Harvard University Press.

Fetal head size in monkeys and apes:

Schultz, A. (1949). Sex differences in the pelves of primates. *American Journal of Physical Anthropology*, 7: 401–423.

On obstructed labour around the world and through time:

Dolea, C. & AbouZahr, C. (2003). *Global burden of obstructed labour in the year 2000. Evidence and Information for Policy.* Geneva: World Health Organization.

Maharaj, D. (2010). Assessing cephalopelvic disproportion: back to the basics. *Obstetrical & Gynecological Survey*, 65: 387–395.

Ould El Joud, D., Bouvier-Colle, M.-H., MOMA Group (2001). Dystocia: a study of its frequency and risk factors in seven cities of west Africa. *International Journal of Gynecology & Obstetrics*, 74: 171–178.

Roy, R. P. (2003). A Darwinian view of obstructed labor. *Obstetrics & Gynecology*, 101: 397–401.

Wittman, A. B. & Wall, L. L. (2007). The evolutionary origins of obstructed labor: bipedalism, encephalisation, and the human obstetric dilemma. *Obstetrical & Gynecological Survey*, 62: 739–748.

On the nature of limbs

Budding limbs and growth plates

On cell death in embryology:

Penazola, C. *et al.* (2006). Cell death in development: shaping the embryo. *Histochemistry and Cell Biology*, 126: 149–158.

On cell death in the developing brain:

Azevedo, F. A. C. *et al.* (2009). Equal numbers of neuronal and nonneuronal cells make the human brain an isometrically scaled-up primate brain. *The Journal of Comparative Neurology*, 513: 532–541.

Low, L. K. & Cheng, H.-W. (2006). Axon pruning: an essential step underlying the developmental plasticity of neuronal connections. *Philosophical Transactions of the Royal Society B*, 361: 1531–1544.

The myotome dance

On *Hox* genes and limb development:

Burke, A. C. & Nowicki, J. L. (2001). *Hox* genes and axial specification in vertebrates. *American Zoologist*, 41: 687–697.

Richardson, M. K. *et al.* (2009). Heterochrony in limb evolution: developmental mechanisms and natural selection. *Journal of Experimental Zoology*, 312B: 639–664.

Fins and limbs

Owen, R. (1849). On the Nature of Limbs. A discourse delivered on Friday, February 9, at an evening meeting of the Royal Institution of Great Britain. London: John van Voorst.

On the genetic switch for fingers and toes:

Gilbert, S. F. (2000). Generating the proximal-distal axis of the limb. In: *Developmental Biology* Sinderland, MA: Sinauer Associates.

A Greenland saga

On *Acanthostega*:

Clack, J. A. (2006). The emergence of early tetrapods. *Palaeogeography, Palaeoclimatology, Palaeoecology*, 232: 167–189.

On *Tiktaalik's* pelvis:

Shubin, N. H., Daeschler, E. B., Jenkins, F. A. (2014). Pelvic girdle and fin of *Tiktaalik roseae*. *Proceedings of the National Academy of Sciences*, 111: 893–899.

Hip to toe

Two legs good

On the position of the foramen magnum as a guide to bipedality:

Russo, G. A. & Kirk, E. C. (2013). Foramen magnum position in bipedal mammals. *Journal of Human Evolution*, 65: 656–670.

Lucy's hips and Johanson's knee

On hominin hips:

Lovejoy, C. O. (2005). The natural history of human gait and posture Part 1: spine and pelvis. *Gait and posture*, 21: 95–112.

On the bicondylar angle of australopithecines:

Lovejoy, C. O. (2007). The natural history of human gait and posture Part 3: the knee. *Gait and posture*, 25: 325–341.

The limitations of human feet

On incredible ankle-bending in tree-climbing people:

Venkataraman, V. V., Kraft, T. S., Dominy, N. J. (2013). Tree climbing and human evolution. *Proceedings of the National Academy of Sciences*, 110: 1237–1242.

On the flexible feet of apes:

Lovejoy, C. O. *et al.* (2009). Combining prehension and propulsion: the foot of *Ardipithecus ramidus*. *Science*, 326: 72, 72e1–72e8.

Putting human feet to the test:

Bates, K. T. *et al.* (2013). The evolution of compliance in the human lateral midfoot. *Philosophical Transactions of the Royal Society B*, 280: 20131818.

DeSilva, J. M. & Gill, S. V. (2013). Brief communication: A midtarsal (midfoot) break in the human foot. *American Journal of Physical Anthropology*, 151: 495–499.

The origins of bipedalism

Stephen Jay Gould on 'The March of Progress':

Gould, S. J. (1989). *Wonderful Life*. New York: WW Norton & Co. pp30–36.

For a flavour of the debate about the origins of bipedalism:

Corruccini, R. S. & McHenry, H. M. (2001). Knuckle-walking hominid ancestors. *Journal of Human Evolution*, 40: 507–511.

Dainton, M. (2001). Did our ancestors knuckle-walk? *Nature*, 410: 324–325.

Kivell, T. L. & Schmidt, D. (2009). Independent evolution of knuckle-walking in African apes shows that humans did not evolve from a knuckle-walking ancestor. *Proceedings of the National Academy of Sciences*, 106: 14241–14246.

Richmond, B. G., Begun, D. R., Strait, D. S. (2001). Origin of human bipedalism: the knuckle-walking hypothesis revisited. *Yearbook of Physical Anthropology*, 44: 70–105.

Upright ancestors as the prevailing view in the early twentieth century:

Avis, V. (1962). Brachiation: the crucial issue for Man's ancestry. *Southwestern Journal of Anthropology*, 18, 119–148.

Keith, A. (1923). Man's posture: its evolution and disorders. *British Medical Journal*, 1: 451–454.

On Ardi's movement in the trees:

Crompton, R. W., Sellers, W. I., Thorpe, S. K. S. (2010). Arboreality, terrestriality and bipedalism. *Philosophical Transactions of the Royal Society of London B*, 365: 3301–3314.

Lovejoy, C. O., McCollum, M. A. (2010). Spinopelvic pathways to bipedality: why no hominids ever relied on a bent-hip-bent-knee gait. *Philosophical Transactions of the Royal Society of London B*, 365: 3289–3299.

On the positions and movements of apes in trees:

Hunt, K. D. (1991). Positional behaviour in the Hominoidea. *International Journal of Primatology*, 12: 95–118.

Thorpe, S. K. S. & Crompton, R. H. (2006). Orangutan positional behaviour and the nature of arboreal locomotion in Hominoidea. *American Journal of Physical Anthropology*, 131: 384–401.

On the relatively late expansion of savannah habitat in Africa:

deMenocal, P. B. (2004). African climate change and faunal evolution during the Pliocene-Pleistocene. *Earth and Planetary Science Letters*, 220: 3–24.

On the anatomy and behaviour of *Oreopithecus*:

Crompton, R. H., Vereecke, E. E., Thorpe, S. K. S. (2008). Locomotion and posture from the common hominoid ancestor to fully modern hominins with special reference to the common panin/hominin ancestor. *Journal of Anatomy*, 212: 501–543.

On climbing and arm-hanging as sensible options for a large-bodied ape:

Larson, S. G. (1998). Parallel evolution in the hominoid trunk and forelimb. *Evolutionary Anthropology*, 6: 87–99.

Born to run

On the wide range of variation in fossils from the site of Dmanisi in the Republic of Georgia:

Lordkipanidze, D. *et al.* (2013). A complete skull from Dmanisi, Georgia, and the evolutionary biology of early *Homo. Science*, 342: 326–331.

On long legs as a possible consequence simply of increased body size:

Pontzer, H. (2012). Ecological energetics in early *Homo. Current Anthropology*, 53, S6: S346–S358.

On Nariokotome Boy's environment – the expansion of the African grasslands:

deMenocal, P. B. (2004). African climate change and faunal evolution during the Pliocene-Pleistocene. *Earth and Planetary Science Letters*, 220: 3–24.

On the evidence for hunting by ancient hominins:

Ferraro, J. V. *et al.* (2013). Earliest archaeological evidence of persistent hominin carnivory. *PLOS One* 8: e62174.

A new kind of novelty

On the malleability of anatomy and how an animal might change to suit its environment before any genetic changes take place:

West-Eberhard, M. J. (2005). Phoenotypic accommodation: adaptive innovation due to developmental plasticity. *Journal of Experimental Zoology*, 304B: 610–618.

On epigenetics and the potential for adaptive changes occurring during an animal's lifetime to be passed on to the next generation:

Bateson, P. (2012). The impact of the organism on its descendants. *Genetics Research International*: 640612.

Shoulders and thumbs

Hominin shoulders

On the late appearance of truly human-like shoulders:

Larson, S. G. *et al.* (2007). *Homo floresiensis* and the evolution of the hominin shoulder. *Journal of Human Evolution*, 53: 718–731.

On the release of stored elastic energy in throwing:

Roach, N. T. *et al.* (2013). Elastic energy storage in the shoulder and the evolution of high-speed throwing in *Homo. Nature*, 498: 483–486.

On the ancient obsidian points from Gademotta:

Sahle, Y. *et al.* (2013). Earliest stone-tipped projectiles from the Ethiopian Rift date to >279,000 years ago. *PLOS One* 8: e78092.

Ape elbows, units and hands

On the shape of elbows in chimpanzees:

Larson, S. G. (1998). Parallel evolution in the hominoid trunk and forelimb. *Evolutionary Anthropology*, 6: 87–99.

On the differences between human and chimpanzee hands:

Marzke, M. W. (1997). Precision grips, hand morphology, and tools. *American Journal of Physical Anthropology*, 102: 91–110.

Marzke, M. W. & Marzke, R. F. (2000). Evolution of the human hand: approaches to acquiring, analysing and interpreting the anatomical evidence. *Journal of Anatomy*, 197: 121–140.

The hands of our ancestors

On curvature in finger bones reducing strain in hands used for hanging from branches:

Richmond, B. G. (2007). Biomechanics of phalangeal curvature. *Journal of Human Evolution*, 53: 678–690.

On Ardi's hand and the independent evolution of long metacarpals in chimpanzees and gorillas:

Lovejoy, C. O. *et al.* (2009). Careful climbing in the Miocene: the forelimbs of *Ardipithecus ramidus* and humans are primitive. *Science*, 326: 70.

On Don Johanson's intrepretation of Lucy's arms and curved fingers:

Wong, K. (2012). Lucy's Baby. *Scientific American*, 22: 4–11.

bigthink.com/videos/what-lucy-looked-like

On flexible spines and abandoning life in the trees:

Lovejoy, C. O. (2005). The natural history of human gait and posture Part 1: spine and pelvis. *Gait and posture*, 21: 95–112.

For more on the hands of *Australopithecus sediba:*

Kivell, T. L. *et al.* (2011). *Australopithecus sediba* hand demonstrates mosaic evolution of locomotor and manipulative abilities. *Science*, 333: 1411.

Unique thumbs

Diogo, R., Richmond, B. G., Wood, B. (2012) Evolution and homologies of primate and modern human hand and forearm muscles, with notes on thumb movements and tool use. *Journal of Human Evolution*, 63: 64–78.

Williams, E. M., Gordon, A. D., Richmond, B. G. (2012). Hand pressure distribution during Oldowan stone tool production. *Journal of Human Evolution*, 62: 520–532.

The making of us

The unlikeliness of being

For more on embryology generally, and estimated numbers of early miscarriages:

Sadler, T. W. (2012). *Langman's Medical Embryology*. Philadelphia: Lippincott, Williams and Wilkins.

On constraints limiting the possible paths that evolution can take:

Muller, G. B. & Newman, S. A. (2005). The innovation triad: an EvoDevo agenda. *Journal of Experimental Zoology B*, 304B: 487–503.

Masters of our own destiny

For a brilliant exposition of the way animals influence their environments:
Dawkins, R. (1982). *The Extended Phenotype*. Oxford University Press.

Human uniqueness

On looking for human uniqueness in the right places:
Cartmill, M. (1990) Human uniqueness and theoretical content in paleoanthropology. *International Journal of Primatology*, 11: 173–192.
Roberts, A. M. & Thorpe, S. K. S. (2014). Challenges to human uniqueness: bipedalism, birth and brains. *Journal of Zoology* (in press).
On why we shouldn't look for a single 'adaptive shift' to explain all the changes in human evolution:
Cartmill, M. (2012). Primate origins, human origins, and the end of higher taxa. *Evolutionary Anthropology*, 21: 208–220.

LIST OF
ILLUSTRATIONS
&
INDEX

ILLUSTRATIONS

INDEX